The Clock Repairer's Manual

THE
CLOCK REPAIRER'S
MANUAL

~ Mick Watters ~

CROWOOD

First Published in 1996 by
The Crowood Press Ltd
Ramsbury, Marlborough
Wiltshire SN8 2HR

British Library Cataloguing-in-Publication Data
A catalogue record for this publication is available from the British Library.

ISBN 1 85223 960 3

Photographs and line drawings by the author.

Typeset and designed by
D & N Publishing
Lambourn Woodlands
Hungerford, Berkshire

Typefaces used: text, M Plantin; labels, Gill Sans.

Printed and bound at The Bath Press, Bath

Dedication
To my wife Coleen, who typed the manuscript for the book, and to my supportive family and friends who have never offered a word of doubt that the work would be completed.

CONTENTS

ACKNOWLEDGEMENTS

I thank Cynthia Boult who was consulted over the cover, colleagues at St Loye's College Foundation, particularly Peter Mitchell and Mike Western for their advice and support on technical matters, and to the trainees of St Loye's College through whom I have learned so much.

PREFACE

This book was written primarily for the horological enthusiast interested in taking that first step into clock repairing and for those under initial training in clock repairing. There is likely to be material which will benefit the professional repairer as well. For example I have never read in any book how to splice a twelve strand long case clock rope. It is included here.

Those completely new to clock repairing will be led gently into new skills and knowledge. It is the intention that this book will inform about techniques and give you the confidence to have a go. It is also the intention that those with limited experience will extend their ambitions.

You will find here a mix of imperial and metric units reflecting the speed of change within industry generally. For example, steel continues to be supplied in imperial units so don't be surprised if you are asked to select ⅛inch brass and set up in a Number 32, (3.2mm), collet.

1 GETTING STARTED

Two early considerations for those preparing to repair clocks from home, whether as pure enthusiast or professional, are where to work and what tools will be needed to make a modest start.

Where to work will include alternatives such as a shed, an outhouse, a loft conversion, a utility room or a garage; otherwise an existing workshop might be converted, or the lounge, dining-room, kitchen or even the bedroom utilized. Perhaps initially working in a bedroom doesn't seem realistic, but in fact making or converting a double wardrobe into a work station is very sensible. There are significant advantages to working in a bedroom: there is minimal disturbance from other users of the house, and work in progress may be left from one day to another – yet to anyone entering, the work station would appear to be an ordinary double wardrobe.

Certainly working inside the house offers the advantage of comparative warmth in the winter months. The disadvantage of working in a lounge, dining-room or kitchen is having to clear things away even as work progresses.

Next to be considered are the workbench, storage space, light and a chair to sit on.

THE WORKBENCH

Many hours are likely to be spent sitting at the workbench, so it must be the right size and shape for your comfort and health, and strong enough for the work that is to be undertaken. It should enable you to work occasionally with both elbows resting on its surface, and with an upright back.

A good height for a clock repairer's bench is about 92cm (an approximate conversion from the traditional imperial height of 36in). A minimum width should also be about 92cm (36in), though if room permits, a bench 107cm (42in) wide would allow a small vice to be permanently attached to the bench top at one end; it would then be available for immediate use but without getting in the way during other bench work. The depth of the bench should be a minimum of 46cm (18in), but 61cm (24in) if space permits; this will allow for the work in hand and offer room for various tools that will be needed as work progresses.

The bench top should be about 2.5cm (1in) thick, or more if possible, to take light hammering and heavy clocks. The surface should be covered with plain brown or light green cork lino or a plain vinyl; it is better to make the surface as plain as possible so that small parts can be seen easily on the surface. The sides and back of the bench should have high edging to contain items on the bench, and the front should have a low edging, for comfort yet also to prevent items from rolling off.

Small drawers under the bench will help with storage for tools and materials. The storage should be well off the floor as it is essential to be able to sweep under the bench to find small parts that will inevitably find their way there. Every potential clock repairer should be aware that small clock components can spontaneously grow both wings and legs and will discover the most inaccessible places to hide in.

The bench should be well lit, with either a twin balanced fluorescent tube lamp or a tungsten lamp with a 40 watt pearl bulb. A stronger bulb will have the disadvantage of throwing out too much heat. Some repairers advocate working by a natural north light, but this is far from essential. Working with sunlight falling directly on the bench is very uncomfortable, not so much because of the heat but because of the uncomfortable light. A blind should overcome the problem of direct sunlight falling on the bench.

Finally a working surface, often of plain white A4 paper, is secured to the bench in front of the repairer.

If the luxury of a purpose-built, dedicated bench is not possible, a table top will do, although raising the working level with blocks or a false top will be more comfortable. The surface of a table top will need protecting to avoid scratches.

A further space with a working surface will be necessary for clock cleaning and other dirty or heavy work. Certainly an outhouse or garage is better suited to some work, but is not essential for a modest start.

Working heights vary so a chair that can be adjusted between about 46cm and 61cm (18 and 24in) should ensure a comfortable working position. A back to the chair is preferable.

TOOLS

For anyone just beginning in clock repairing, the thought of a significant outlay on tools would perhaps be off-putting. In fact much can be achieved with ordinary tools found in the home. Inevitably as knowledge and skills are gained, the desire for specialist tools will increase; but these can be acquired over a long period, perhaps being paid for by repairs carried out.

Tools need not be new, of course, and the repairer is encouraged to attend some of the many horological tool auctions held across the country. British Horological Institute branches often hold local auctions which members of the public may attend; BHI branch secretaries will have advance notice of pending auctions, and their addresses may be obtained from the secretary of the BHI.

Fig 1 An ideal clock repairer's bench.

Fig 2 A clock repairer's bench laid out with the tools in constant daily use.

The absolute minimum needed to get started will be small flat nose pliers, a variety of small screwdrivers, one or two small plastic containers to hold parts, and a pair of tweezers. For those who wish to purchase a more comprehensive range of specialist tools the following are recommended, and should cost no more than the equipment needed for many other activities such as archery, fishing, mountain biking or photography. These may be purchased from tool and material houses or from second-hand sources.

Tool List
Tweezers, general purpose
Set of watchmaker's screwdrivers, 3mm to 1mm
Two oil cups
Essence jar, 60mm (2.5in)
Brush, four-row
Movement blower
Bench knife
Material tray
Clock hammer, 50g
Screwdriver, 180mm × 5mm
Set of four oilers
Eyeglass, single
Eyeglass, double × 10
   (spectacle type for those who prefer)
Pliers, flat-nose 115mm
Pliers, round-nose 115mm

Top cutters flush 115mm
File, 150mm (6in) No4 double cut
File handle

Other tools for various repairs will be identified as they become necessary.

Tool Care
Now a few words about the care of tools. Those not in constant use should be kept close by in drawers, leaving the bench free to carry out the work in hand. Tools that need special care include tweezers, screwdrivers and oilers; these are also the most used and abused.

Tweezers are used more than any other tool, and if properly looked after will save you having to spend hours searching for parts which otherwise will fly out of them. Use them for dismantling, assembly and balance spring adjusting. Avoid levering and prising with them, and any activity that might bend them. When the points are squeezed together, they should remain in contact for some distance back from the point; if they are bent out or in, any parts you are attempting to hold will in fact fly out of them. Tweezers are dressed by grinding and filing roughly to shape, then emery cloth is used round the outside faces to give them a pleasing appearance. Avoid touching the inside faces altogether.

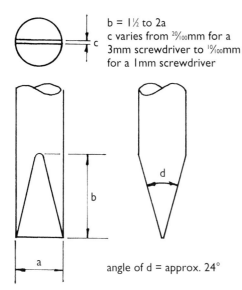

b = 1½ to 2a
c varies from $^{20}/_{1000}$mm for a
3mm screwdriver to $^{10}/_{1000}$mm
for a 1mm screwdriver

angle of d = approx. 24°

Fig 3 A well-proportioned screwdriver.

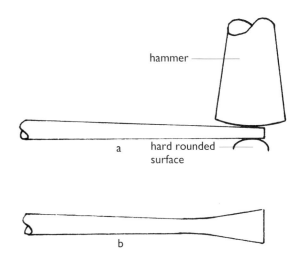

hammer

hard rounded surface

Fig 4 The oiler filed to a gentle taper (a) and flattened (b).

The best screwdrivers are hardened and tempered to dark blue. Screws used in horology, even with the same tap size, often have differing width slots so it is necessary to sharpen screwdrivers with this in mind. To sharpen, the screwdriver is placed in the bench vice and filed to the proportions shown in Fig 3. Use a 'broken in' 150mm No 4 double cut file for this, or similar. Small screwdrivers may be sharpened on an India stone lubricated with clock oil.

MAKING OILERS

Oilers can be purchased already well shaped and in different sizes, but inevitably they are eventually damaged. The following notes on making an oiler can also be used for repairing a damaged one:

1. Select a piece of 'blue steel' 50mm (2in) long and a diameter to suit the size of oiler being made. Alternatively, select a sewing needle and temper the pointed end to a light blue for one third of its length.
2. File the steel to a gentle taper, then rest the filed end on a hardened rounded surface, for example the handle of a pair of top cutters, and flatten the end with a light hammer.
3. Stone the two flats so that they are uniform then stone the sides to the required shape. Stoning across the face of the oiler gives greater control over the way the oil leaves the oiler.
4. Make a handle from an old plastic knitting needle, using techniques adapted from the section on repivoting. When the blade is inserted, the oiler is ready for use.

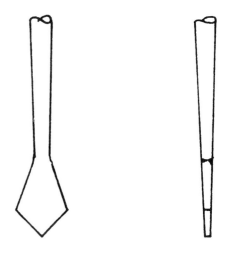

Fig 5 The finished oiler.

2 THE CLOCK MOVEMENT

Before stripping your first clock you should know the preferred name of some of the components common to most mechanical clocks, and some of the definitions, so that the instructions given are better understood. Every activity has its own 'language', and communication is easier if that language is understood (or at least enough to get by and build on).

Fig 6 A striking clock movement.

Even the word 'clock' has different meanings for different people. For some it is the case, dial, hands and all the things that can be seen from the outside; to others it is what can't be seen, that is, the 'workings'. In horology, once the 'workings' have been separated from the case, dial and hands it is known as 'the movement'.

The movement comprises a number of components, each of which will come under a certain heading. For example, every movement will have some sort of framework to hold everything together. There will be a driving force to keep the clock going, usually a weight or spring, and a series of wheels of the right ratio to make the hands go around at the right rate and for the right duration. There will be some means for winding a spring or weight without having to turn all the wheels backwards, some means of adjusting the hands independently and, very importantly, some means of releasing the power from the mainspring or weight under control. These mechanisms will now be looked at more closely.

CLOCK TRAINS

To start with then, our movement has two plates: a bottom plate sometimes called a pillar plate or a dial plate, and a top plate sometimes called the back plate. Between the plates there is a series of wheels called a train which usually, though not always, gives a step up in gear ratio from where the drive originates. In an alarm clock there will be two trains, a time train and an alarm train. In a striking clock there will also be two trains, a time train and a striking train. Similarly a chiming clock will have three trains, a time, strike and chime train.

The time train in a simple thirty-hour clock (the name given to a clock that is wound daily) usually consists of a great wheel, which is the first wheel in the clock; a centre wheel, which is the second wheel in the clock; a third wheel and a fourth wheel. The latter may have a long pivot to carry a seconds hand.

The Escapement

Next is the escapement, which allows the power from the mainspring or weight to escape slowly under control. In a pendulum clock there are two main parts to the escapement, an escape wheel and a pallet. In a clock with a balance there are three main parts, an escape wheel, a pallet and a balance.

Sometimes there is another wheel between the great wheel and the centre wheel called the intermediate wheel; this converts the clock to an eight-day clock. When an intermediate wheel is used, the wheel driven by the centre wheel is still often called the third wheel, even though it is actually the fourth wheel in the train. The only time that this may be a problem is when ordering such a new wheel from a material house. Describing its position in the train should overcome any ambiguity.

The great wheel always has some arrangement to hold the mainspring in a wound state; usually this incorporates a ratchet wheel, click and click spring. The centre wheel will have a simple clutch arrangement of some kind to allow the hands to be turned independently of the rest of the train.

The prime function of the escape wheel is to give impulse to a pendulum or balance to keep it swinging. How escapements work will be looked at later.

A balance will vibrate according to the natural laws of springs, and a pendulum will vibrate according to the law of gravity. It should be appreciated that all time-measuring devices involve an 'event' which is repeated in the same space of time over and over again and is counted up. This applies to the earth as a clock, a water clock, a sundial, a pendulum clock, a clock working with a balance or even a quartz clock.

Fig 7 The clutch arrangement on a centre wheel to allow handsetting.

Motion Work

On the outside of the front plate will be found another series of wheels, this time called 'motion work'. Motion work makes use of a cannon pinion, a minute wheel and pinion, and an hour wheel. The motion work gives the 12:1 reduction to drive the hour wheel. A 24-hour dial will have a 24:1 reduction.

Earlier, when talking about the time train, reference was made to a wheel. The word 'wheel' is often used to mean the whole wheel which is usually made of two separate parts fastened together, one part a wheel, the other a pinion.

In horology a wheel is a driver: it is made of brass and has teeth. A pinion is driven, and is made of steel and has leaves. The teeth of the wheel drive the leaves of the pinion. There are exceptions to this, for example the cannon pinion in the motion work; however, it is true to say that generally, pinions make bad drivers.

Endshake and Sideshake

A wheel must have up and down movement between the plates, and this movement is called

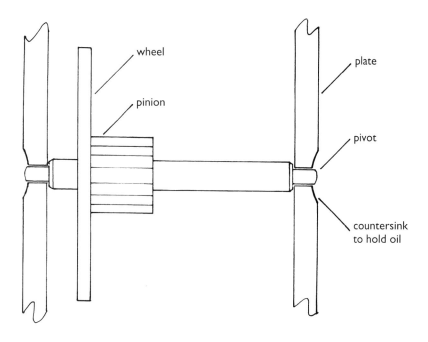

Fig 8 A typical wheel and pinion in a plate.

endshake. There must also be clearance between the sides of the pivots and the plates, and the movement of the pivot due to this clearance is called sideshake. Endshake must always be a minimum of $\frac{3}{100}$mm in any wheel or staff; in a long-case clock, it may be $\frac{1}{2}$mm or more, though it need not be as much as this.

Safe sideshake is indicated by a 10° lean in all directions when a wheel is placed in a plate and allowed to lean naturally.

SIMPLE TRAIN CALCULATIONS

For most repair work, the only calculations that need to be carried out are those for sorting out invoices, mark-ups and profit margins. However, occasionally you may need to calculate the number of vibrations per hour or per minute or even per second that a clock has to make in order to keep perfect time, or you might have to calculate the correct ratio of teeth to leaves of a missing wheel in a clock so that you can arrive at the most likely combination for a new wheel and pinion to be cut.

Calculating Vibrations

When calculating the number of vibrations a pendulum or a balance needs to make per hour, per minute or per second for the clock to keep perfect time, first we need to know the number of teeth used in an hour on one of the time-train wheels. The calculation then becomes the product of the drivers multiplied by two, divided by the product of the driven to give the number of vibrations per hour. If the number of vibrations per minute or second is required, we simply divide by sixty once or twice, to get minutes or seconds.

As most clocks – though not all – have a centre wheel, we take that as a starting point because we know that every tooth will be used once in an hour. (The centre wheel rotates once in an hour, therefore every tooth will be used just once.) Our formula for calculating the number of vibrations per hour looks like this:

$$\frac{t\ c/w \times t\ 3rd \times t\ 4th \times t\ scape \times 2}{l\ 3rd \times l\ 4th \times l\ scape} = vib/hour$$

t = number of teeth   l = number of leaves
c/w = centre wheel

The times two in the top line is because the escape wheel advances half a tooth at a time, and the centre wheel pinion doesn't enter the calculation because it comes before the centre wheel, as does the great wheel.

As a practical example, a train count reveals the following numbers:

c/w	54 teeth		
3rd wheel	40 teeth	3rd pinion	6 leaves
4th wheel	40 teeth	4th pinion	6 leaves
scape wheel	15 teeth	scape pinion	6 leaves

I recommend that the formula you are going to use is always written down first.

$$\frac{\text{t c/w} \times \text{t 3rd} \times \text{t 4th} \times \text{t scape} \times 2}{\text{l 3rd} \times \text{l 4th} \times \text{l scape}} = \text{vib/hour}$$

$$\frac{54 \times 40 \times 40 \ 15 \times 2}{6 \times 6 \times 6} = 12000 \text{ vib/hour}$$

If vibration per minute is wanted, just put '× 60' in the bottom line, or divide the answer by 60. For vibration per second, just put '× 60 × 60' in the bottom line or divide your vib/hour calculation by 3,600 (60 × 60). Should you need the time of one vibration, simply divide your vibrations per second into the number 'one'. In other words, the time of vibration is the reciprocal of the number of vibrations per second.

Very occasionally a clock comes in for repair with a missing wheel, and a calculation has to be made to determine the likely number of teeth in the missing wheel and the likely number of leaves in the pinion. Provided you know the number of vibrations per second, minute or hour that the pendulum or balance makes for the clock to keep time, then it is relatively straightforward. For example, if the missing wheel and pinion was from a long-case clock with a second pendulum, you would know that the pendulum vibrates sixty times a minute. If you were unsure, you could even count the number of vibrations in a minute with a stopwatch. It would bring you so close to

a convenient ratio that the correct number of teeth and leaves would be determined.

The following is the train count of a long-case clock with a missing third wheel. Calculate the likely number of teeth in the wheel and leaves in the pinion.

t c/w 64	
t 3rd (unknown)	l 3rd (unknown)
t scape 30	l scape 8

Number of vibrations per minute = 60

Note that there is no fourth wheel. In fact it doesn't matter, just leave it out of the calculation; it still works out right.

When doing a train count, use two pairs of tweezers: hold the wheel on a tooth with one pair of tweezers while you touch each tooth in turn with the other pair as you count. It is a good idea to count twice to make sure that the first count was the correct number.

The formula this time is:

$$\frac{\text{t c/w} \times \text{t 3rd} \times \text{t scape} \times 2}{\text{l 3rd} \times \text{l scape} \times 60} = 60 \text{ vib/min}$$

Notice the 60 in the bottom line: that is because we are using the formula for calculating the number of vibrations per minute. Now, substituting numbers for our formula and calling the unknown teeth in the wheel 'A' and the leaves 'B', we have:

$$\frac{60 \times A \times 30 \times 2}{B \times 8} = 60 \text{ vib/min}$$

Bring the unknown to one side of the formula, and the known to the other side, and we end up with:

$$\frac{A}{B} = \frac{60 \times 30 \times 2}{8 \times 60}$$

$$= \frac{15}{2}$$

This means that the ratio of wheel to pinion is 15:2.

Now certainly the wheel would not have had 15 teeth and I don't recall ever seeing a pinion in horology of under 6 leaves, so that rules out the possibility of a 30-tooth wheel and a 4-leaf pinion. Try multiplying 15 and 2 by 3. That brings us to the possibility of a 45-tooth wheel and a 6-leaf pinion. It would probably work, although experience tells me that a higher number of teeth would have been used in this type of clock. So next I'm looking at a 60-tooth wheel and an 8-leaf pinion, and this sounds very likely, in fact more likely than 75 teeth for the wheel and 10 leaves in the pinion.

Probably all I need to do now is supply my gear cutter with the distance between centres of the centre and third wheel, and the distance between centres of the third wheel and scape wheel together with the number of teeth and leaves and the outside diameter of both centre wheel and scape pinion, and a new wheel could be cut. The distance between the plates would also be required.

CALCULATION FOR MOTION WORK

The ratio between a cannon pinion and an hour wheel is 12:1; the formula for calculating motion work of a twelve-hour dial is therefore as this:

$$\frac{\text{t minute wheel} \times \text{t hour wheel}}{\text{l cannon pinion} \times \text{l minute pinion}} = \frac{12}{1}$$

Supposing the hour wheel is missing and the number of teeth is to be determined. The solution is:

teeth minute	leaves cannon
wheel 36	pinion 36
teeth hour	leaves minute
wheel unknown	pinion 36

$$\frac{\text{t minute wheel} \times \text{t hour wheel}}{\text{l cannon pinion} \times \text{l minute pinion}} = \frac{12}{1}$$

$$\frac{36 \times A}{36 \times 6} = \frac{12}{1}$$

$$A = 12 \times 6$$

By calculation, the teeth in the hour wheel (A) are 72.

A calculation for the hour wheel or cannon pinion will always result in the actual number of teeth or leaves being discovered. Calculations for missing minute wheels will normally end with a ratio. The long-case clock is an exception because usually the ratio between cannon pinion and minute wheel is 1:1; that is, there are normally the same number of teeth in the minute wheel as there are teeth in the cannon pinion. (The cannon pinion in a long-case clock is an exception to the rule of pinions having leaves and wheels having teeth.)

3 CLEANING A SIMPLE CLOCK MOVEMENT

A 'timepiece' is the usual name given to a clock without strike or chime or alarm. A timepiece is likely to have two plates, one train (a time train), an escapement and motion work. There will be a ratchet arrangement to hold a mainspring or weight in a wound state, and a clutch arrangement to turn the hands independently. A pendulum or balance may be used to control the speed of the running down of the train.

There is a wide choice of clock for our first practice, but a Smiths eight-day timepiece has been chosen for three reasons: such clocks are still around in reasonably large numbers and might be acquired through car boot sales quite cheaply; they are representative of many other clocks, and so principles can be grasped and transferred; and finally, they are simple.

The first thing to do when tackling a repair is to establish why a clock has stopped. Often the reason is simply that the old oil has dried and congealed, causing a drag on the train and escapement. This condition is easily recognized by looking at the countersinks around the pivot holes and observing the old, dried, congealed oil. The condition is often accompanied by a more generally dirty-looking movement, particularly around the pinion leaves and the other working surfaces.

Pivot holes may be worn, causing inefficient transmission of power. Gripping the great wheel between index finger and thumb and reversing the normal direction of drive of the train usually shows this up by causing pivots to jump across worn holes. Worn holes often appear with what looks like black, congealed oil.

Steel pivots may be cut because foreign bodies are embedded in the pores of brass pivot holes; this condition is often associated with red dust around the pivot and hole. It can be dealt with in a number of ways (these will be looked at later).

Attempting to wind the clock will show up a broken mainspring. If no power can be felt at all during winding, probably the mainspring is broken at, or near its inner end. When some power can be felt during winding but the mainspring can be wound indefinitely, probably the spring is broken at, or near its outer end.

Repairs will be covered in subsequent chapters; for the moment we shall concentrate on dismantling, cleaning, reassembling and oiling our simple eight-day timepiece, and the principles explained here are likely to hold true for any mechanical spring-driven clock. The detail will need to be modified if a different clock is being tackled.

DISMANTLING

1. First, remove the pendulum, which simply lifts off.
2. Remove the clock hands. There are two common ways of securing clock hands, and these are shown in Figs 9 (a) and (b). In the first, the tapered securing pin is removed using flat-nose pliers: rest one side of the pliers against the thinner end of the pin, and the other side against the centre wheel arbor; then squeeze. Assuming the pin loosens, change the grip to the thicker end of the taper pin, and pull it out. There should be a washer which will lift off, then the minute hand should be eased off. The hour hand can now be removed by holding the boss in the centre between thumb and index finger, twisting slightly and pulling.

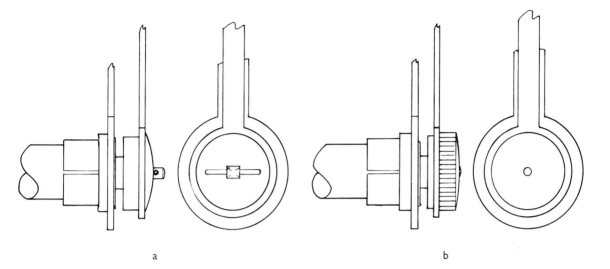

a   b

Fig 9 Two common ways of securing clock hands: (a) is pinned; (b) is screwed.

Fig 10 The suspension spring of a typical mass-produced clock.

In the second way, as illustrated in Fig 9b, unscrew the finger nut with thumb and index finger, and lift the hands off.

3. Remove the movement from the case. This is usually a simple task involving unscrewing six or eight wood screws which are accessible through the case back. Lift out the movement.

The pendulum and the pendulum rod are suspended by the suspension spring; in this type of clock the latter is usually a flexible piece of spring with brass cheeks top and bottom.

4. Remove the suspension spring and suspension rod; these are removed first because the suspension spring is vulnerable. Draw out the securing pin from the pallet cock and pull the suspension spring downwards slightly to clear the cock. Turn the suspension rod through 90° and lift it off the crutch.

5. Next the power must be removed from the mainspring: it is essential to do this before stripping any clock. The operation has to be carried out with great care to avoid damaging the ratchet wheel teeth and to avoid getting a nasty clout from the winding key. In the absence of a letting-down key, use the clock winding key to let the mainspring down.

Letting the Mainspring Down

(a) Hold the clock on your lap, and fit the winding key on the winding square.

(b) Wind the ratchet wheel by half a tooth, then with a piece of pegwood, disengage the click by pressing on its tail. The click needs to clear the ratchet wheel, but by *very little* if damage to the click spring is to be avoided. Holding the click in this position, let the mainspring down by half a turn, then release the click and wind by one tooth. Check that the click is fully engaged.

(c) Without releasing your grip on the key, lift it off the winding square; then with a twist of the wrist, reposition it on the winding square again and repeat the operation.

Keep this up until all the power is removed from the mainspring; and when you feel you have reached this point, check by feeling that the great wheel is free. It must be sufficiently free of power to be rocked easily in a clockwise and anti-clockwise direction within the limits allowed by a tooth in between two leaves of the intermediate wheel pinion.

Every clock repairer has experienced a knock from an uncontrolled key but strict adherence to the above will minimize the risk. A letting-down key has no wings and is smooth so it can be allowed to turn under control in the palm of the hand while the click is held clear of the ratchet wheel (Fig 11).

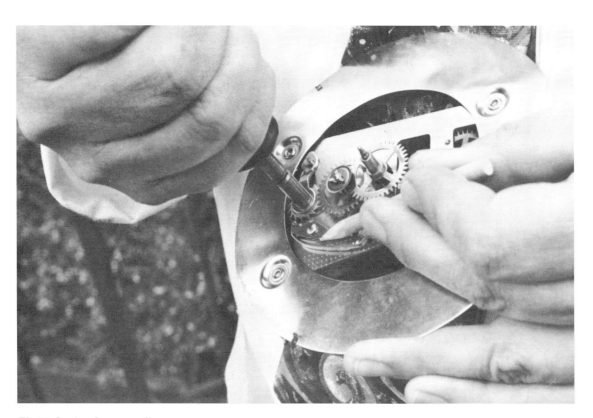

Fig 11 Letting the power off.

6. Remove the two screws holding the pallet cock, the pallet cock itself and then the pallets. There may be washers under the pallet-cock screws.

7. Remove the minute-wheel retaining clip (often a straight taper pin or a wrap-around pin), the washer, the hour wheel and the minute wheel.

8. Unscrew the bridge holding the ratchet wheel and remove the ratchet wheel. The click and click spring may be riveted to the plate (as they are here), or they may be screwed. Remove the screws as well.

There are three common methods of holding the plates of clocks together, and in each instance the pillars, usually four, may be riveted into the other plate. Both top and bottom plates may be fastened in the same manner.

9. Remove the more convenient plate by first unscrewing the four pillar nuts. If the front plate has brackets to secure the movement to the case, try not to move these; instead remove the back plate, carefully avoiding bending any pivots. To help achieve this, lift the plate vertically, keeping the plates parallel.

10. Remove the intermediate wheel, the fourth wheel in the train and the escape wheel by lifting them vertically from the plate to avoid bending a pivot. Often the barrel housing the

mainspring can also be removed at this stage, but the centre wheel cannot be removed without first removing the cannon pinion. (It may be that the barrel cannot be removed until the centre wheel has been driven out.)

11. Remove the cannon pinion. Support the plate with a stout watch-brush handle (otherwise an old file covered with paper to protect the plate will suffice), and drive the centre wheel out with a mallet, holding the clock over the bench. An old clock hand

Fig 12 Removing the cannon pinion (right).

Fig13 Three ways of securing clock plates (below).

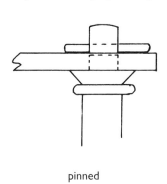

pinned

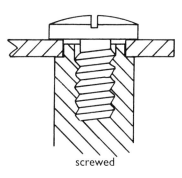

screwed

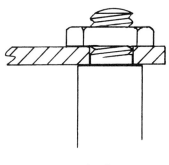

secured with nut

placed on the square of the centre wheel will prevent the centre arbor splitting the mallet or getting embedded into the mallet. Make sure the driving blow is straight to avoid bending the arbor, although a slight bend can be straightened later. The barrel will fall out during this operation unless previously removed.

12. The barrel usually has a cut-out in the cover intended to assist cover removal. Mark the wall of the barrel to identify the relationship between the cover and the wall; in quality clocks this is often done by a light countersink in the wall beside the cut-out.

13. Remove the barrel cover by holding the wall of the barrel under the teeth in one hand and striking the end of the barrel arbor with a mallet. Usually this doesn't dish the cover, but if it does, the dish is easily taken out. Removing the cover by making use of the cut-out provided often damages it, and the method explained is the better way.

Taking Out a Dish in a Barrel Cover

Support the outside edge of the cover on an old barrel, and either press on the centre with a press tool, or use a punch on the centre of the cover; flatten by striking the punch with a mallet or hammer.

14. Remove the barrel arbor by gripping the square with pliers, and turning the arbor in the opposite direction to winding and drawing out slightly. If necessary, lift the inner coil of the mainspring away from the arbor with a thin screwdriver. Hold the barrel in the opposite hand for this.

15. Next, the mainspring should be removed from the barrel, in particular to inspect the outer end to ensure that it isn't torn. To remove the mainspring, in the absence of a mainspring winder, hold the barrel in a gloved hand and grip the mainspring with long-nose pliers by an inner coil, then carefully lift out just one or two coils to begin with. Gradually ease out the remaining coils under control. This is not an easy task, and

you should take care to point the barrel away from your face in case the mainspring comes out suddenly. Often the last coil has to be unhooked from the wall of the barrel. Once the mainspring is out of the barrel, check the end. Look for tears in the steel between the hole (eye) and the mainspring end.

The dismantled clock is now ready for cleaning.

CLOCK CLEANING

Although clock cleaning can be mechanized, cleaning by hand is still widely practised. The process usually involves a cleaning solution, a compatible rinsing solution and some form of drying. The cleaning solution should loosen congealed oil, assisted by a scrubbing action with a stiff bristle brush, while the rinsing solution must dry without leaving deposits on the clock.

Proprietary clock-cleaning and rinsing fluids are available through material houses. Some are toxic, others are non-toxic. There are waterless fluids and concentrates to be diluted with water, there are ammoniated and non-ammoniated cleaning fluids, fluids that will brighten dull brass and fluids for use in an ultrasonic cleaning machine. Which sort is used on clocks will depend on the method of cleaning – whether by hand, by ordinary machine or ultrasonic cleaning machine – and also on which cleaning fluid your supplier stocks, and your willingness to compromise on appearance for the sake of conservation.

Traditionally clocks were cleaned in a mixture of ammonia and soft soap, perhaps with other additives, then rinsed and dried. Great emphasis was placed (or misplaced) on creating a polished finish, so the clock was returned to the customer fully restored but perhaps with some unwitting damage to the original maker's finish. For a long time this was of no concern to the practical man, but it is now an issue. Some repairers still use a buffing machine to polish plates and wheels, but not only does this tend to destroy the original finish on brasswork, it may also tear holes.

Contemporary horologists are beginning to differentiate between repair, restoration and conservation, so more care is being taken to retain as much of the original clock as is reasonably possible, including original surface finishes. Proprietary cleaning fluids with a high concentration of ammonia are not recommended for clock cleaning as the ammonia can cause damage to brass plates, wheels and other brass parts. What is recommended as a compromise is brushing in a cleaning fluid with a low concentration of ammonia, then further brushing with a compatible rinsing fluid, followed by rapid drying.

Drying cleaned parts is usually effected by one of the following methods:

(a) immersion in boxwood dust (from the material house);
(b) wiping with a cotton cloth;
(c) blown warm air (perhaps with a hair dryer).

Whether you are enthusiast or professional, cleaning a clock in paraffin is worthy of consideration. Its advantages include low cost, ease of purchase, relatively high flash-point and low toxicity. The clock is simply scrubbed with a stiff bristle brush and dried thoroughly with a non-fluffy cloth. Paraffin will not brighten the brasswork as proprietary cleaners may do, but neither does it dull bright brass nor does it cause lacquer to lift. In addition to this I know of no damage caused to brass by paraffin.

Safety Recommendations
Making up your own cleaning fluid is not recommended as you might end up with an unstable fluid which could be a health risk to yourself and bad for the clock. If a toxic proprietary brand of fluid is used, take notice of any safety recommendations by the maker, but in addition, take any extra precautions which seem reasonable. For example:

• Avoid using in a confined space.
• Have adequate ventilation.
• Use an extractor fan if available.
• Never smoke whilst using cleaning fluids.

• Avoid naked flames.
• Know what to do in case of accident.
• Label storage cans.
• Avoid skin contact if you have a skin complaint; use gloves if necessary.
• Never tip toxic fluids down the sink or drain: it is illegal and dangerous. Instead, arrange disposal through your local authority
• Don't store large quantities of toxic or flammable fluids in the immediate working area.

If making use of proprietary cleaners, cover the parts being cleaned to ensure that a 'tide mark' is not left on brasswork between what was in, and what was out of the fluid. Prolonged soaking in some cleaning fluids will brighten dull brass, but beware of soaking for too long for fear of etching it. Avoid soaking the clock in fresh cleaning fluid for more than half an hour.

Once the clock has been cleaned and thoroughly dried, by whatever means, the pivot holes are pegged clean with pegwood. Peg both plates from both sides to remove any congealed oil still remaining after the cleaning process. Also, inspect the pinion leaves to ensure that no dirt remains between them, that they have not been cut because of the action of the wheel and pinion, and that there is no rust. Examine the pivots to ensure that they are straight and polished, with a square shoulder. Finally, check that there are no bent teeth in the barrel or train wheels.

REASSEMBLY

As assembly work progresses, it will be necessary to lubricate the clock with special horological lubricant; these are readily available from material houses. Three grades will be required: heavy, for lubricating clock mainsprings and other relatively high pressure points; medium, for lighter-loaded train-wheel pivots; and light, for platform escapements.

1. Start by replacing the mainspring. Grip the bottom of the barrel in the left hand, for a right-handed person, with the open end fac-

ing up. Now hold the mainspring in the right hand, such that it can be fed over the hook in the barrel. Feed more of it into the barrel, making sure that the second coil passes the barrel hook. As you feed in the rest of it be sure to keep distortion to a minimum. Gloves will help to protect your hands and will prevent sweat being transferred to the spring which could be a cause of breakage.

2. Replace the barrel arbor, making sure that the mainspring is properly hooked.

3. Lubricate the mainspring with a heavy grade of clock oil. The oil should be sufficient to give the whole spring a thin film, but not so much that it runs out of the barrel.

4. Replace the barrel cover, having lined up the previously positioned marks, by clamping in a soft jaw vice. Use paper to protect the barrel and cover from marking. Make sure the barrel cover is replaced the correct way up.

5. Check the endshake of the barrel arbor which should be in the order of about $^{10}/_{100}$ to $^{20}/_{100}$ mm of a millimetre, though it is often unnecessarily more.

6. Lubricate the barrel arbor pivoting points which engage with the barrel with heavy clock oil. Put the barrel to one side until required.

7. The centre wheel has a clutch arrangement that allows hand adjustment. Hold the brass wheel while turning the steel arbor, and observe the points where friction occur. Lubricate those points with heavy oil – they are likely to be between the arbor and pinion and between the tension spring and the pin holding the spring in place. Check that there is sufficient friction between the arbor and the wheel so that the hands will carry: this is a particularly important check. Adjust if insufficient.

8. Replace the barrel, intermediate wheel, centre wheel, fourth wheel and escape wheel into the bottom plate.

9. Position the top plate over the pillars and transfer your grip on the movement so that the thumb on your left hand is under the bottom plate and your fingers are holding the top plate down but under light pressure only. Now, with your tweezers, manipulate each pivot into its hole, starting with the barrel and finishing with the escape wheel.

10. Replace the four nuts which hold the plate on, tightening them securely.

11. Test the freedom of the train by holding the movement horizontally, then lifting each wheel in turn, ensuring that they drop under their own weight. Turn the movement over and repeat the operation: if a wheel doesn't lift or drop under its own weight, investigate and correct. Look in particular for a bent pivot or distorted plate: each would need correcting.

OILING

Avoid over-oiling. Correct oiling will be indicated by a small fillet of oil between the shoulder of a pivot and the plate when the wheel is lifted so that the shoulder contacts the plate.

1. Lubricate the barrel, the intermediate wheel and the centre wheel top and bottom pivots with heavy clock oil.

2. Lubricate the fourth and escape wheel top and bottom pivots with a medium grade of clock oil.

3. Replace the cannon pinion as follows: support from beneath the centre wheel back pivot with a flat hollow punch; this supports the brass plate around the pivot, and does not bear on the pivot itself. Now, with another flat hollow punch, drive the cannon pinion to a position where it is just tight on the arbor and no more. Never drive the cannon pinion down to touch the plate: it will probably cut all power from the train.

Temporarily replace the hour wheel, and check that the shoulder on the centre wheel which supports the minute hand just shows through the hour pipe. This will ensure that when the minute hand is fitted there will be endshake on the hour wheel.

4. Oil the underneath side of the ratchet wheel with heavy oil, and replace over the square on the barrel arbor.

5. Replace the ratchet wheel bridge after oiling the top face of the ratchet wheel, and secure with its screw. Make sure the correct screw is used: bad marking is often seen on the barrel, and this is caused by too long a screw being used.

6. Oil the minute wheel post with medium oil, and replace the minute wheel.

7. Replace the hour wheel.

8. Replace the minute-wheel washer and clip, checking endshake of the minute wheel after securing.

9. Replace the pallets, pallet cock and pallet cock screws, also the washers if present, and depth the pallets, by moving the pallet cock up or down, to give a safe drop, nominally 1 degree. It will be necessary to put a little power on the mainspring for this (see the next chapter for setting up the escapement).

10. Replace the suspension spring and suspension rod thus: feed the rod over the crutch, pass the suspension spring between the cheeks of the block on the pallet cock, and pin. After pinning, the suspension spring must be free but with no play. Tightness here is a cause of stoppage.

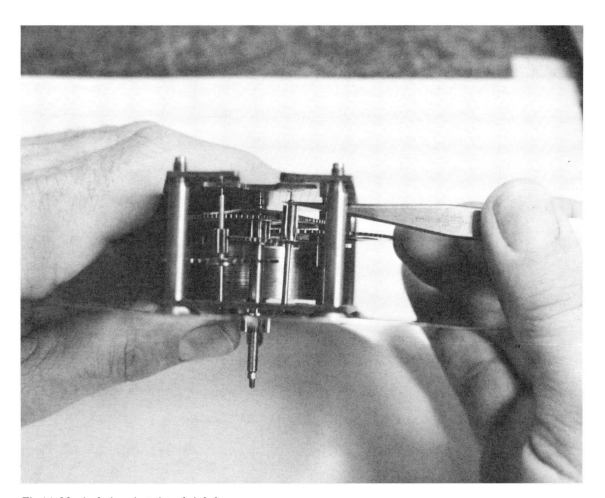

Fig 14 Manipulating pivots into their holes.

If there is too much play, close the cheeks of the suspension block slightly; if too tight, either open the cheeks slightly or dress off a little of the brass sides of the suspension spring. Ideally the cheeks of the block should be parallel.

11. Oil with light oil the escape wheel pivots, the pallet faces and escape wheel teeth, and put a little on the crutch where it engages with the suspension rod. This last lubrication point is necessary because there is friction between the crutch and suspension rod due to the pallet and suspension moving about different centres. There must always be a small amount of play between the crutch and suspension rod: if there is not, and if there is a lack of lubrication at this point, it can cause the clock to stop.

ADJUSTING THE MOVEMENT

Having given the whole movement a final check, replace it into the previously cleaned case, then secure it with six or eight screws; replace the hands. Ensure that there is endshake in the hour wheel after securing the minute hand.

Finally, check the clock for timekeeping. An inexpensive mass-produced clock of this type might be expected to keep time to within a couple of minutes a week. When regulating an eight-day clock, adjust the rate daily when the errors are large, but once timekeeping is within about half a minute a day, leave the clock to run for a full seven days so that it can be regulated according to the average daily change.

Lowering the bob will cause a comparative loss, raising it will cause a comparative gain (the amount of correction necessary for a given gain or loss can easily be calculated using the formula provided earlier). As an approximation, one turn of the rating nut on a mass produced eight-day clock similar to the one described here will give a difference of about two-and-a-half minutes a day, depending on the pitch of the thread in the rating nut. One turn of the rating nut on a long-case clock with a seconds pendulum will give approximately one-and-a-half minutes a day change.

4 THE PENDULUM AND ASSOCIATED ESCAPEMENTS

The purpose of explaining how escapements work is primarily to enable you to set them up competently yourself, but also to help with fault diagnosis. Fundamentally, what is required of our escapement is to allow the power of the mainspring or weight to be released slowly and under control and to give periodic impulses to the pendulum to keep it swinging. The pendulum will swing according to the relatively stable laws of gravity, and consequently our clock has the potential to keep time.

THE PENDULUM

A pendulum will have some form of suspension, most often a strip of steel spring; a rod, which is commonly of iron, steel, brass, wood or invar; a bob which is often lead, brass or wood, as in the case of a cuckoo clock; and a rating nut, usually of brass, to alter the length of the pendulum. From my own early days I recall the definition of a pendulum as being 'an object free to swing about an axis not passing through its centre of gravity'. Clearly a clock pendulum matches the definition.

When a pendulum is hanging vertically and stationary it is said to be 'in the position of rest'; When it is swinging from side to side it is 'vibrating'. In horology, unlike in physics, one vibration is taken to be the swing of a pendulum (or balance) from maximum displacement on one side of the position of rest, through the position of rest, to maximum displacement on the other side. The angle through which the pendulum swings is called 'the arc of vibration', and the time taken for one vibration is called 'the time of vibration'. The 'amplitude' of a pendulum or balance, which we sometimes measure, is the

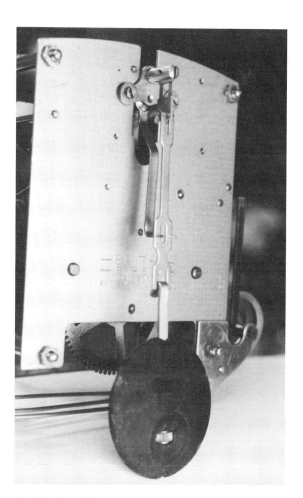

Fig 15 A pendulum.

angle from the position of rest to maximum displacement on one side.

The pendulum can therefore control the running down of a mainspring or weight by receiving impulse, usually with each swing; and because the time taken for each swing is very nearly constant, the mechanism that delivers the impulse can be fitted with the means to count the number of swings of the pendulum – although instead of literally counting swings, it is calibrated to show hours, minutes and seconds, or any other practical division of time. Day, date, month and moon phase are typical examples of other measurements of time.

Pendulums can be 'real' or 'imaginary'. The 'real' pendulum is the pendulum found on a clock and is sometimes referred to as a 'compound' pendulum. The 'imaginary' pendulum, sometimes called a 'simple' or 'theoretical' pendulum, does not and cannot exist because it is based on a heavy weight suspended by a weightless thread of constant length; however, it is used for the purpose of calculations, and comparisons can be made to produce a real pendulum.

The point of flexure of the real pendulum is in the suspension spring, just below its point of suspension; this corresponds to the point from which the imaginary pendulum swings. The centre of gravity of the real pendulum lies just above the centre of the bob; this corresponds to the centre of the bob on our imaginary pendulum. From this it may be seen that by calculating the length of a theoretical pendulum, the approximate length for a real pendulum could be determined.

Calculating Pendulum Length

The formula for calculating pendulum length is:

$$L = \frac{t^2\, g}{\pi^2}$$

where **L** is the theoretical length of the pendulum expressed in metres, **t** is the time of one vibration or one swing, **g** is the acceleration due to gravity which may be taken as 9.81m/s^2, and π – strictly $^{22}/_7$ – may be taken as 3.14. These values are close enough for the practical purpose of replacing a missing pendulum. Once the calculation is done, a replacement pendulum is provided based on experience of the type of pendulum used on a particular type of clock.

Factors Affecting the Rate of a Clock

There are other factors which affect the rate of a clock, one particularly worthy of note being the stiffness of the suspension spring. I have seen a number of clocks with the pendulum almost touching the bottom of the case, yet the clock still gained. In each instance, a weaker suspension spring reversed the gain and the pendulum rate was brought under the control of the rating nut in the usual way.

It is worth noting that increasing or decreasing the weight of the pendulum bob has no effect on the rate – that is, by adding weight at the centre of gravity of the bob. If weight is added above the centre of gravity, it would have the effect of shortening the pendulum; if it is added below the centre of gravity, it would have the effect of lengthening it. Ordinarily in repair work, however, the weight of a pendulum is not tampered with.

A further point to note is that a one-second pendulum is four times longer than a half-second pendulum.

Much has been written about the time a pendulum takes to swing through a small arc compared with the time it takes to swing through a large arc, yet a distinction is not always made between free-swinging pendulums, and the effect of the escapement on a pendulum when it is in a clock. Certainly it would be desirable to have varying arcs of vibration performed in the same length of time, a condition described as **isochronism,** but contemporary pendulum clocks in particular are designed to give a near-constant impulse combined with a low arc of vibration to the pendulum. Recoil escapements often have arcs of about 5°, sometimes a little more; deadbeat escapements are designed to have an arc of about 3° when small variations in arc have minimal effect on timekeeping.

The difference between the time taken to describe a small arc and that for a large arc is known as 'circular error'.

Pendulums swing in a circular path. If that path could be made cycloidal, isochronism would be achieved but devices to achieve this, for example cycloidal cheeks, have proved unsuccessful.

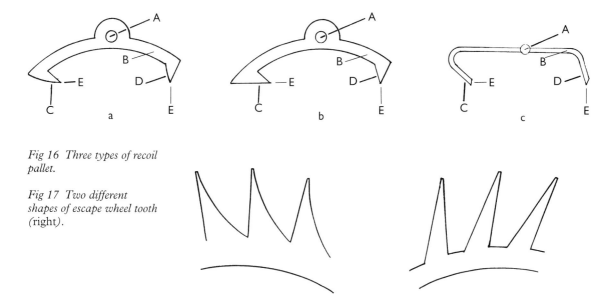

Fig 16 Three types of recoil pallet.

Fig 17 Two different shapes of escape wheel tooth (right).

THE ANCHOR OR RECOIL ESCAPEMENT

Impulse is given by the escape wheel to the pallet which is secured to the pallet arbor. The pallet arbor carries a crutch so the impulse is passed to the pendulum rod to keep the pendulum swinging. The 'action' between the escape wheel and pallet is fairly simple geometry. A study of the following is recommended to set pallets up correctly: follow it through step by step, and it should be easy to understand the action. Try to study the stages rather than concentrate on the whole. The example used is typical of the pallet and escape wheel found in long-case clocks, but it will be found elsewhere in other forms.

The pallet, or pallets as they are often called, are made of steel, hardened dead hard on the working faces. The shape of recoil pallets can vary as in Fig 16; in each case 'A' is the pallet centre, 'B' is the pallet arm, 'C' is the impulse face of the entry pallet, 'D' is the impulse face of the exit pallet and 'E' is the discharging corners of both pallet faces. The escape wheel which works with these pallets can also vary in shape.

The Action of the Recoil Escapement

Impulse
Imagine the pendulum to be at maximum displacement on one side of the position of rest with power on the escape wheel trying to drive it in a clockwise direction. Gravity will pull the pendulum towards the position of rest, but at the same time the escape wheel, through tooth (A), will give impulse to the left-hand pallet impulse face. (The left-hand pallet nib is called the entry pallet, because it is the first of the two pallet nibs visited by a tooth.) At the end of the impulse the tooth is at the discharging corner of the entry pallet (Fig 19).

Drop
Tooth (A) leaves the entry pallet freeing the escape wheel completely until tooth (B) is arrested by the other pallet nib, called the exit pallet. The action of the escape wheel turning freely under this condition is called 'drop' and amounts to a nominal 1°.

Recoil
Due to inertia, the pendulum will swing just a little further after drop which causes the impulse face

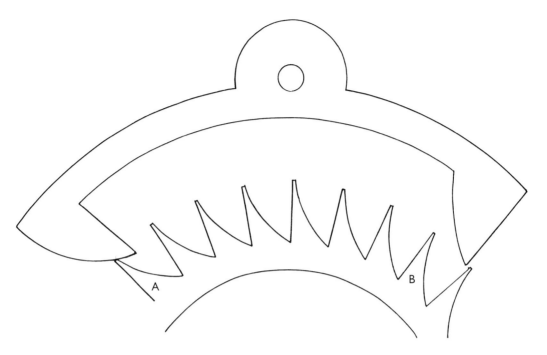

Fig 18 The start of impulse on the entry pallet.

Fig 19 Impulse is over when the escape wheel tooth is at the discharging corner of the pallet.

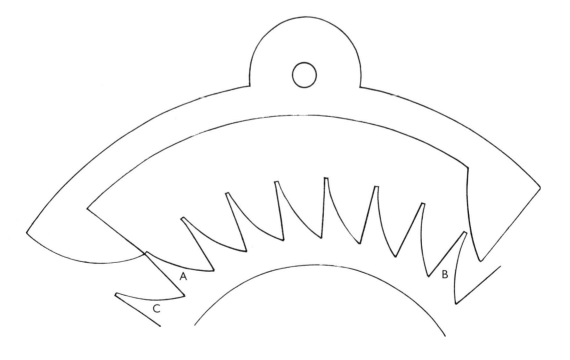

of the exit pallet to push the escape wheel backwards slightly, often by 2° or 3°. Hence the name recoil escapement. This explains why the seconds hand of a long case clock moves backwards by about half a second after each 'tick' and 'tock'. (The tick and tock are, of course, the sound of the drop on the entrance and exit pallet.)

Eventually the pendulum will stop swinging momentarily, then gravity will pull the pendulum back towards the position of rest. Immediately the pendulum starts to fall, impulse will be given by tooth (B) sliding along the exit pallet impulse face.

As soon as tooth (B) is at the discharging corner of the exit pallet, the escape wheel is free to turn again and so drop occurs once more. Tooth (C) lands on the impulse face of the entry pallet but once again, the pendulum over-swings slightly, again making the entry pallet push the escape wheel backwards slightly.

We have now seen a full cycle, and this action is repeated until the clock runs down: impulse, drop, recoil; impulse, drop, recoil...

THE DEAD BEAT ESCAPEMENT

The dead beat escapement was invented in 1715 by the famous maker George Graham, and in spite of being a frictional rest escapement, it is the type used in many of the best timekeepers. It is similar to the recoil escapement in many ways but a significant difference is that the pallet faces are two distinct faces, a locking face and an impulse face, meeting at the locking corner. Also the locking faces are arcs of circles, though not the same circle, struck from the pallet centre.

The Action of the Dead Beat Escapement

Unlocking
We start, as usual, with the pendulum at maximum displacement on one side of the position of rest. The pendulum falls due to gravity and the locking face of the entrance pallet sweeps across an escape-wheel tooth tip until eventually the tooth and the locking corner of the pallet become coincidental. The unlocking is now over.

Impulse
Impulse is given as the tooth slides along the impulse face of the pallet, and is complete when the tooth meets the discharging corner of the pallet. Impulse is usually 2°.

Drop
The escape wheel is now briefly free of the pallets, and turns through a nominal 1° until another

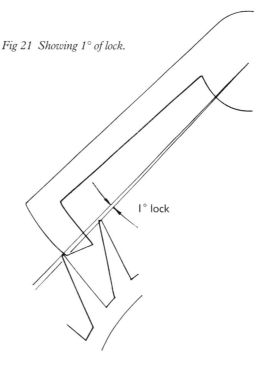

Fig 21 *Showing 1° of lock.*

1° lock

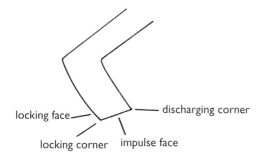

locking face — discharging corner

locking corner   impulse face

Fig 20 *The names of the acting surfaces of a dead beat pallet.*

tooth drops onto the locking face of the exit pallet. The point of contact between the tooth tip and the locking face should be 1° up from the locking corner, measured from the pallet centre.

'Lock' is the name given to the depth of engagement between the tooth and pallet. It is 1° as described above.

Rest

The pendulum doesn't suddenly stop after lock but, due to inertia, swings further by about a degree. The escape wheel remains stationary on the locking face of the pallet, but there is friction between the pallet and scape tooth as the pallet is still moving. Hence, frictional rest escapement.

Eventually the pendulum stops, then due to gravity, starts to fall as before. This time the tooth unlocks from the exit pallet and gives impulse to the exit pallet, then when it leaves the discharging corner of the exit pallet, again drop occurs

and continues for 1° turn of the escape wheel when another tooth drops onto the locking face of the entry pallet again.

THE BROCOT ESCAPEMENT

This is a very attractive escapement when fitted outside the dial on a French strike. The pallet faces are semicircular, and were usually made of cornelian, kept in position with shellac. Unfortunately when a pallet stone becomes worn or broken, the only practical solution is to make a new 'stone' out of steel and it never looks so pleasing, although it does get the escapement going again.

The escapement is not truly a dead beat, but it is close to it and is often referred to as a half dead beat escapement. When correctly set up – which they often are not – at the moment of drop, the tip of an escape wheel tooth should drop onto the entrance

Fig 22 A Brocot escapement.

Fig 23 The relationship between the pallet and escape wheel on completion of drop.

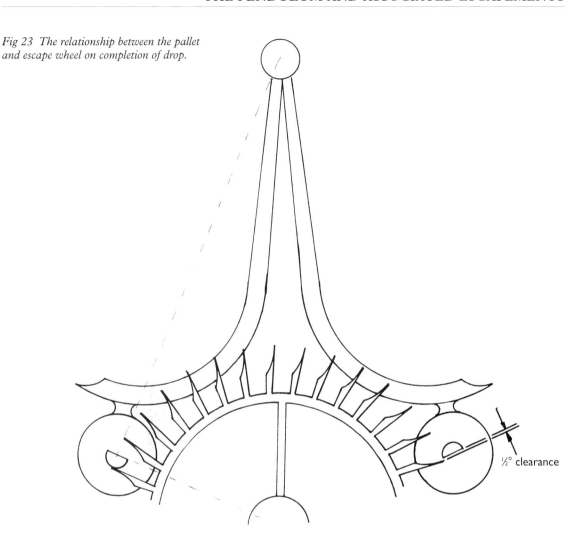

½° clearance

pallet on the centre line, or slightly above but not below, joining the pallet stone and the pallet pivot.

Action of the Brocot Escapement

Yet again, the action starts with the pendulum at maximum displacement on one side of the position of rest.

Impulse

The moment the tooth passes the centre line of the pallet and pallet centre, impulse is given and continues until the escape wheel tooth meets the discharging corner of the stone.

Drop

The escape wheel turns through approximately 1°, then another tooth drops onto the exit pallet with the tip of the tooth contacting the stone at, or slightly above, its centre.

Lock

Once the tooth drops onto the exit pallet, the escape wheel is locked; but due to the slight over-swing of the pendulum, the escape wheel may be seen to advance a little further if the leading faces of the teeth are radial and the amplitude of the pendulum is high.

Once the pendulum has completed its swing, it comes back again, unlocking the escape wheel which gives impulse to the exit pallet stone. When the tooth reaches the discharging corner of the stone, the escape wheel is free again and so drop onto the other pallet stone occurs once more.

Problems Encountered with the Brocot Escapement

Perhaps a lot of the problems with this escapement occur for three reasons. First, when it is fitted to a French clock with visible escapement, there is an eccentric bush for the front pallet pivot which invites adjustment. Second, the pallet arms are split, again inviting adjustments. Third, the pallets are secured with shellac, which can come away, necessitating resetting.

Should the escapement be badly adjusted, one or more of the following contingencies may have to be effected or taken into account:

1. When properly adjusted, the flats on both pallets should be at right-angles to the imaginary line joining the pallet stone centre and the pallet pivot centre.
2. The eccentric bush for the pallet front pivot may be very tight and will be prone to seizure. Run a drop of oil around the eccentric bush before turning it. If the slot is chewed before you start, the bush is probably already seized. Knock it out, lubricate it and replace it, and then adjustments may be made once again.
3. It may be necessary to open or close the pallet arms to correct previous attempts at adjustment.
4. The pallet stones may need uprighting.
5. The pallets or escape wheel may need bushing to prevent the teeth catching intermittently on pallets.

A particularly difficult and frustrating job can be uprighting pallets that are set in holes too large for them. When making new pallet pads, turn them out of blue steel, if necessary with a step, so that the working faces are of the correct diameter and the end of the pads that fit into the pallet arms are an interference fit.

The semicircular working part should have a diameter slightly less than the distance between two teeth. If excessively less, drop will be at the expense of impulse and the clock may not work.

ADDITIONAL INFORMATION

- The part of the action when a tooth drops onto the entry pallet is called 'outside drop'. 'Inside drop' occurs when a tooth drops onto the impulse face of the exit pallet. Theoretical drops are 1° and should be the same on both pallets.
- Although the drop is 1°, because the escape-wheel teeth are normally ½° thick, the leading edge of an escape-wheel tooth will be 1° in advance of the discharging corner of a pallet after drop, but there will be only ½° clearance between the discharging corner of a pallet and the escape tooth.
- Drop is necessary to ensure that the scape wheel advances, but it should be minimal, consistent with safety allowing for wear, sideshake of both scape and pallet pivots in their respective holes, and the almost inevitable slight uneven spacing between scape teeth. (NOTE: 'scape' and 'escape' are interchangeable.)
- Many clocks have steady pins in the pallet cock, and so the pallets cannot be adjusted easily. Mass-produced clocks often do not have steady pins, so the depth of the pallets have to be set. The pallet-cock screw-holes in such clocks are usually oversize or elongated to allow depthing.
- When depthing recoil pallets, adjust for 1° of drop by raising or lowering the pallet cock, keeping it parallel with the plates and avoiding lifting or lowering just one side. As 1° may be difficult to judge, remember, after drop the clearance between the departing tooth and the pallet discharging corner will be equal to about the thickness of an escape-wheel tooth tip.
- Increasing or decreasing the distance of centres between escape wheel and pallet will affect both drops, but one drop will be affected more than the other. Try to achieve 1° of drop on

both pallets, then check the action as all teeth pass both pallets.

- When depthing dead beat escapements, depth to .1° of lock, then check that the drop is safe all the way round the wheel. The Brocot escapement is depthed in a similar way to the recoil escapement in that the pallets are depthed to give a safe drop of 1°, then the drop is checked all the way round the wheel.

- The number of escape-wheel teeth varies between about thirty and forty. The design data for a clock will include the number of teeth embraced by the pallet. This is always stated assuming a tooth has just dropped onto the entry pallet and will include, with the number of teeth that can be counted between the two pallets, a half tooth. So, commonly pallets will embrace six-and-a-half, seven-and-a-half or eight-and-a-half teeth.

- Suppose the arc through which a pendulum swings while receiving impulse is 4°. The actual total arc will be 4° of impulse, *plus* a little bit of over-swing after impulse (due to the inertia of the pendulum), giving a total arc of about 5°.

A larger arc is undesirable, as small variations in arc when the total arc exceeds 3° do show up in timekeeping.

BEAT SETTING FOR PENDULUM CLOCKS

Understand this section and you should be able to set up a pendulum clock without having to prop up one side of a mantel clock, or any other clock to get it going, or to follow pencil lines on the wall to get a wall clock going. The term 'in beat' will be explained, as will how to carry out the beat test and how to correct a clock that is 'out of beat'.

Definition of 'In Beat'

A clock is said to be 'in beat' when the pallets operate symmetrically about an imaginary straight line joining the pallet and escape-wheel centres, while the pendulum operates symmetri-

cally about a vertical line. This can be both for a clock with a pallet planted vertically over the escape wheel, and for one with a pallet offset.

To enable beat setting to be carried out safely – safe for the clock, that is – there may be a clutch arrangement associated with the crutch and pallet arbor, or possibly, as in the case of a long-case clock, the crutch itself may have to be bent. Some arrangements will require the pallets to be supported while the angle between pallets and crutch is changed, whilst others have some built-in facility to protect the escape wheel from the pallet during beat setting.

Fig 24 Long-case pallets. Support the pallet and bend the crutch to put in beat.

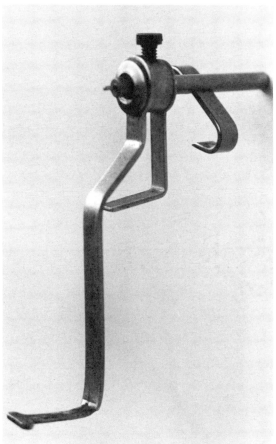

Fig 26 Vienna regulator. Turn the thumb screw to adjust the beat.

Fig 25 A mass-produced strike. Beat setting is effected by moving the crutch to the appropriate side until a resistance is felt, then pushed a little further.

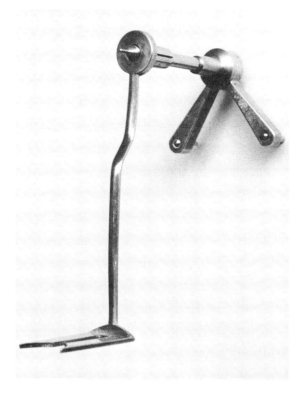

Fig 27 French strike. Often the crutch is screwed to a tapered pallet arbor by a split collet. Approximate beat adjustment is achieved by giving the pendulum a generous swing. Fine adjustment is made by turning the bezel, and therefore the movement, in the case.

Testing for Beat

One simple test for beat is to listen to the clock to see if the 'tick' sounds even. If it does, go on to the fine adjustment test. If it doesn't, lift one side of the clock case to see if that makes the tick more even. If it is worse, lift the other side. Lifting one or other of the two sides should make the tick sound more even.

For approximate beat setting and to get an even tick, move the crutch towards the side that was lifted until a resistance is felt, then move it just a little further. This will alter the angle between the pallet and crutch. Keep this up until an even tick is heard.

Fine Adjustment for Beat

Complete the rough adjustment first if it is possible, though it is not essential, then move the pendulum across to one side of the clock until drop occurs. Immediately it does so, stop moving the pendulum and release it. The clock may stop or continue going; for the moment it doesn't matter. Now do the test again, moving the pendulum to the other side of the clock.

If the clock continues to go from both sides, it is in beat. If the pendulum refuses to swing from one particular side – that is, if drop doesn't occur after releasing it – move it across to that side until a resistance is met, then move the crutch just a little further. The clock is in beat when it continues to work from whichever side the pendulum is released, and it should now stay in beat unless moved to another level.

A wall clock should be levelled with a spirit-level first, then the fine beat adjustment should be carried out.

5 THE WATCHMAKER'S LATHE

A lathe is not essential to basic clock repairing but it is of significant help when polishing pivots, turning collets to mount wheels to arbors and a host of other small jobs. It is virtually essential to have a lathe for repivoting or turning balance staffs. This chapter describes the basic skills needed to operate a watchmaker's lathe competently; how to prepare the graver (the name given to a hand-turning tool); and how to turn a balance staff.

PREPARING THE GRAVER

If you don't already have one, purchase a 2mm to 3mm (1/16 to 1/8in) square graver from a materials house, also a 75mm (3in) file handle, and then fit the graver to the handle by gripping it in the soft jaws of a vice and driving the handle on with a mallet. The handle should be pre-drilled for this.

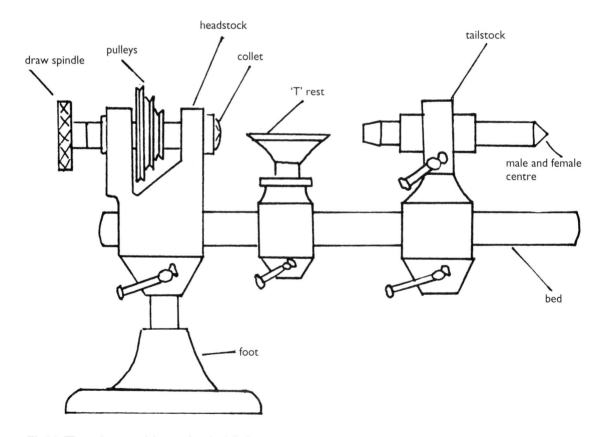

Fig 28 The main parts of the watchmaker's lathe.

Fig 29 A ground general-purpose graver.

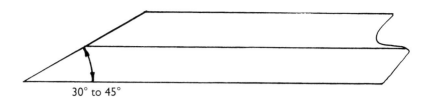

30° to 45°

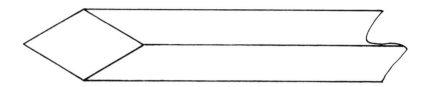

1. Wearing safety glasses, grind your graver to an angle of between 30° and 45°, choosing the sharpest corner for the point. Aim for one uniform face.

2. On an oiled Arkansas stone, stone the two sides of the graver next to the point to remove any burrs. Place the graver across the full width of the stone, keeping it flat and moving it in a circular motion.

3. Grinding will have destroyed the hardness of the graver, so the next step is to bring it to a cherry red colour with a gas torch. This must be done in shaded conditions – it is important that neither sunlight nor any bright light falls on the steel. When the steel is cherry red, continue to apply heat to the end of the graver for two or three seconds: this allows the metal to undergo a structural change. It is only necessary to bring the point and about the first 20mm (¾in) from the point to a cherry red, though keep the flame away from the *very* point to avoid burning it.

4. Now, while the graver is still cherry red, quench it by passing it through oil into water, agitating it in the water to speed the cooling process. Floating oil on the water has the effect of reducing the tendency for the steel to crack.

5. When the steel is cool, remove the graver from the water and oil mix, and dry it.

6. Clean up one side of the graver next to the point on the Arkansas stone or, for quickness, on an abrasive block such as is used for cleaning printed circuit boards; this doesn't seem to round its sharp corners. Clean until the metal is quite bright so that later, tempering colours can be seen.

7. Play the flame on the steel once more 20mm (¾in) from the point, building the heat up slowly and watching for the first colour change, light straw, to creep to its point. Once the point and cutting edges are light straw, quench again either into water, or through oil into water. For this operation the oil is unnecessary.

8. Finally, stone the two sides of the graver next to the point until the metal is bright, then stone the front face of the graver until it, too, is bright. Feel and see that this face is flat on the stone.

The graver should now be ready for use. It is suitable for turning brass, soft steels and harder steels, for example 'blue steel' which has been hardened and tempered to a dark blue colour. It should be possible to leave an almost polished finish on brass or steel with this tool.

As the cutting edges and point become blunt, they may be restored with an India followed by an Arkansas stone without grinding.

TURNING

Eight-millimetre collets normally increase in size in ²/₁₀mm steps, starting with a number 2 (²/₁₀mm) and increasing to about a number 78 (7.8mm). In the smaller sizes, collets are available rising in ¹/₁₀mm sizes, but they are very expensive.

Select 3mm (⅛in) brass rod, 50mm (2in) long, and set it up in a number 32 collet with about a 15mm (⅝in) overhang. Sit in front of, and over the lathe, adjusting the 'T' rest so that it is close to the work and so that the graver can be presented to the work on an imaginary centre line passing through the drawspindle, pulleys and workpiece.

Place your right elbow on the bench, and the fleshy part of the side of your hand on the bed. Now place the graver into your right hand so that the blade is under your thumb, but with the handle to the outside of the palm. Place your index finger close to the point of the graver with the ground end facing up. This is the right-hand grip – although it is possible to turn the lathe round

for left-handed people, I find they have no more difficulty in learning to use a right-handed lathe than right-handed people.

Now place the graver and your index finger on the 'T' rest, and then position your left index finger close to the right so that it is also on the graver and 'T' rest; this is to give support to the graver while turning. The middle finger of your left hand is tucked around the 'T' rest so that the tool is drawn into the work.

Without turning the lathe, practise positioning the graver on the brass: it should be close to the collet with its point below the centre line and not touching the brass, but with its side, just back from the point, on the centre line. Move it across the work to the right, then lift clear. Again, place it on the brass close to the collet, move it across the work to the right, then lift clear. Keep practising this until it feels relatively comfortable and natural.

Wearing safety glasses, turn the lathe on and repeat the above, this time actually taking a cut. If done properly with a sharp graver, the swarf should come away in 'curls'.

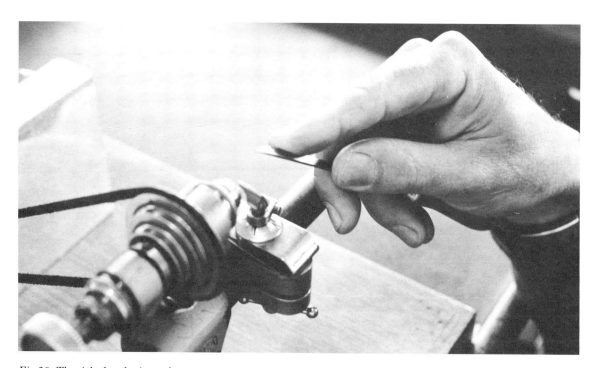

Fig 30 The right-hand grip on the graver.

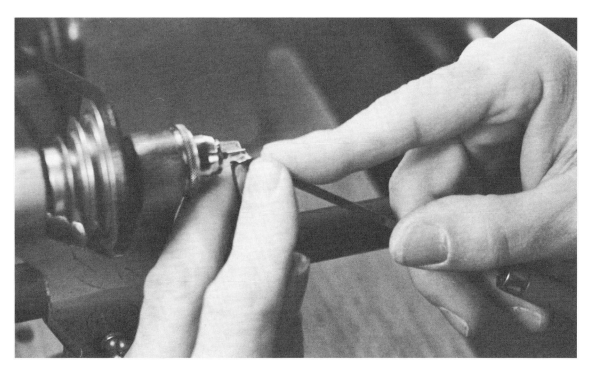

Fig 31 *The complete hold of the graver.*

Fig 32 *Presenting the graver to the work.*

The turning must be under control: never let the work control the tool; the tool must control the work. As you come away from the end of the brass, don't just slip off, but ease the pressure on the tool so that the graver continues to move in a straight line even when it is past the end of the brass. Once you have mastered this very basic use of the lathe, move on to the following exercises, which are the basis of all turning.

NOTE: If you have difficulty seeing what you are doing, either clip an eyeglass over your safety glasses, or pop one lens out of your safety glasses and insert an ordinary eyeglass.

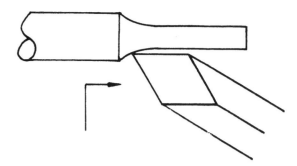

Fig 34 Turning a square shoulder.

Turning Parallel

Swing your right hand down and slightly clockwise, when viewed from the top, placing the whole of the cutting edge of the graver on the centre line of the brass. Now start turning on the left and move to the right as before, remembering to keep the whole of the cutting edge on the brass.

Practise turning the brass to about one half its original diameter, using a new piece as necessary. If it is turned significantly smaller than half its original diameter, it may be that it doesn't cut so well at the right-hand end because of the 'give' in it. Be careful not to turn so thin that the work rolls up over the tool.

Turning a Square Shoulder

Start by turning a piece of brass 3mm (⅛in) in diameter, to approximately half its original diameter. Now use the side and point of the graver and

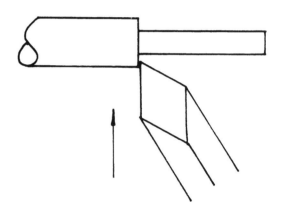

Fig 35 Cleaning up the shoulder.

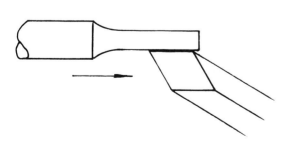

Fig 33 Turning parallel.

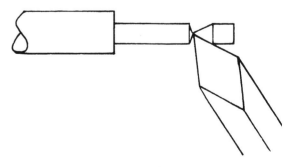

Fig 36 Parting.

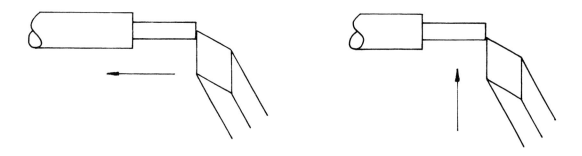

Fig 37 Facing.

begin squaring up the shoulder; next swing the graver further round to tidy up the shoulder, working towards the corner.

It is necessary to be able to achieve a 'dead sharp' corner. For this, a good point on the graver is essential.

PARTING

Often the length of a piece of metal must be reduced, and to do this use the side and point of the graver. The actual angle is not critical.

FACING

To face a piece of metal means to 'square off' the end. Simply use the side and point of the graver in one of the two ways shown in Fig 37. The point of the graver need not pass the centre of the work.

PICKING UP A CENTRE

On occasions, for example pivoting, it is necessary to drill a concentric hole in a piece of metal. Just as a centre punch would be used to locate a drill before drilling, say, a plate, so a similar location must be created in order to concentricly drill a rod. First turn the 'T' rest anti-clockwise 90°, adjusting the height if necessary. Bring the graver to the centre of the turned rod, which has been faced, with the left-hand cutting edge at an angle of 45° to the face; then turn the centre so that it is a 90° included angle, with a diameter a little larger than the web of the drill so the drill can be located.

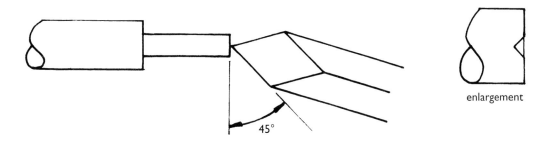

enlargement

45°

Fig 38 Picking up a centre.

It is essential, if a drill is to remain on the centre, that there is no pip at the root of the centre. Now put a drill in a pinchuck, and drill a hole of about 2mm (¹⁄₁₆in) diameter. Keep the drill and pinchuck parallel with the bed of the lathe.

When you are pleased with your efforts and achievements, practise the same exercise on carbon steel.

When turning, remember that collets are notorious for being out of true; in general, therefore, select a piece of metal oversize, turn it true, and then turn it to size, finishing all diameters without removing the metal from the lathe.

The method so far shown of holding the graver is the traditional one. However, some clock repairers use the graver upside-down, much to the horror of the traditionalist. In fact the upside-down method has merit because for some operations the point will last longer. I have used both methods at different times and find that each has its benefits.

An alternative shape graver is also shown, the advantage of this one being that when the cutting edge is damaged, it can be stoned back quickly and put to use again. With the first graver

described, the point soon becomes worn which means having to grind and heat-treat too frequently.

LATHE SET-UPS

When you use a watchmaker's lathe with collets you will soon realize the limitations. Often, particularly when restoring a long-case clock, you will find difficulties in getting the work to turn true, and problems just holding parts in the lathe. These can be overcome by turning between centres using a carrier chuck and carrier at one end of the work and a tailstock with a centre, male or female, at the other end. Alternatively, a tailstock with a lantern runner or jacot drum could be used.

A Set-up Using a Jacot Drum

For polishing pivots using a jacot drum, hold the pivot on the drum with a finger of your left hand while you switch the lathe on. Once the lathe is turning and the burnisher is applied to the pivot, the pivot will be held in place by the burnisher.

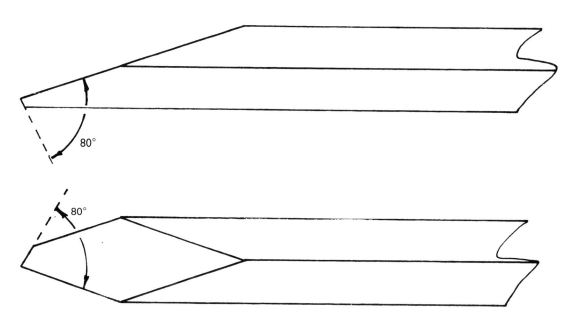

Fig 39 An alternative shape graver.

Fig 40 Use of a carrier and carrier chuck ensure concentricity throughout the wheel.

Fig 41 Making use of a lantern runner.

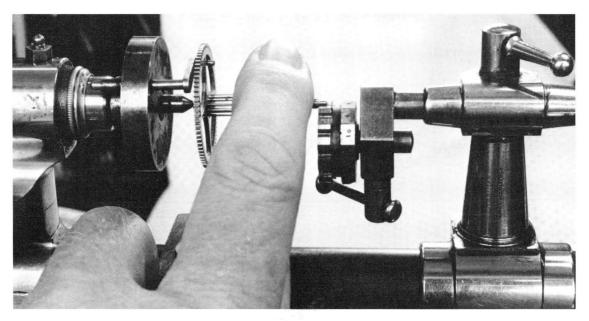

Fig 42 Using a jacot drum.

The beauty of the set-up using the jacot drum is that if something did go wrong and the wheel came out of the lathe, no damage would be done. The arbor could be colleted when using both lantern runner and jacot drum, but never hold the pivot in a collet because it would almost certainly break.

Fig 43 A simple set-up for turning a new arbor.

A Set-Up for Turning a New Arbor

A very simple set-up for turning a new arbor is shown in Fig 43. The arbor can be removed as often as you like and can be turned round, end for end, maintaining concentricity. The steel is first provided with male centres of 60°, and is then put into the lathe so that it is free but with no play. Do not lock the carrier chuck to the carrier even if there is a hole in the carrier for this.

It is important to lubricate the pivot in the centre in the tailstock. As work progresses, heat will be generated as you turn and the arbor will expand, so keep checking the freedom of the arbor. Readjust as necessary.

With care, the pivots of the arbor can be turned between centres, the waste simply broken off and the pivots polished in an ordinary collet.

Set-Up Using a Lantern Runner

When repivoting, holding problems can be overcome with a lantern runner. Hold the wheel so that once more the wheel is free but with no endshake. The end with the broken pivot is held giving access to pick up the centre. To drill, make a spade drill from carbon steel, or make up a holder for a tungsten carbide drill so that it will pass through the hollow centre in the tailstock.

Other Lathe Holding Problems

Some other lathe holding problems can be solved with a three-jaw self-centring chuck with reversible jaws for holding both internally and externally. An independent four-jaw chuck is also useful.

A wax chuck can be used for holding small items such as the endpiece of a platform escapement, or for making a new barrel bush for a long-case clock. In the absence of a proper wax chuck, a piece of brass rod could be held in the lathe, faced to make it true, then the brass could be warmed to take shellac or sealing wax to hold small pieces.

To thin down a strip of brass or steel – for making American clock recoil pallets, for example – you could apply double-sided sticky tape to a face plate in the lathe. It is really quite surprising what can be achieved on such an unsophisticated piece of workshop equipment with a bit of initiative.

Wheel Topping

Before leaving lathe set-ups, there is one last application of the carrier chuck and carrier that is worth showing: it is 'wheel topping'. It is important to maintain a constant impulse to a pair of pallet faces, and it helps if the teeth tips of an

Fig 44 The set-up for repivoting using a lantern runner.

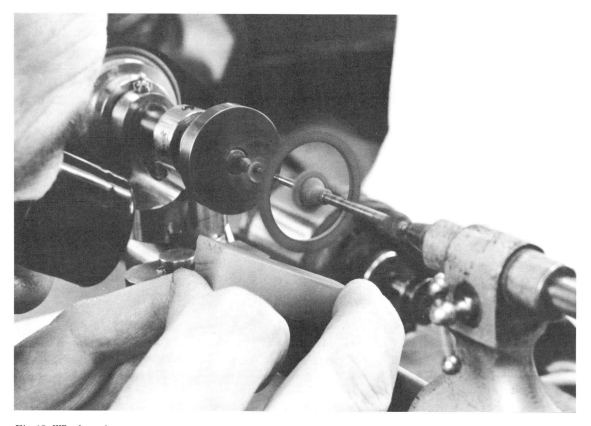

Fig 45 Wheel topping.

escape wheel are all the same distance from the scape-wheel centre.

To top an escape wheel – which is not done automatically, but only when it is felt to be necessary – set the wheel up between centres so that it is free but with no endshake. Lubricate the pivot that revolves in the female centre. Hold an Arkansas stone very steadily under the escape wheel supported by the bed, and allow the teeth of the escape wheel just to touch the stone as the wheel revolves. The object is to reduce the length of the longest teeth until they are the same length as the shortest tooth, reducing the diameter of the escape wheel by the minimum necessary.

When every tooth has been touched, remove the escape wheel from the lathe and smooth any burrs from the teeth with a stone.

After topping the wheel, it may be necessary to adjust the depth of the pallets to accommodate the increased drop caused by slightly reducing the diameter of the escape wheel.

The Compound Slide

A particular advantage of a centre lathe for clock repairing is its power, and anyone who has had to turn a barrel arbor for a chiming or striking clock on a watchmaker's 6mm or 8mm lathe with a graver would agree. However, such work can be made easier with a compound slide where a turning tool is held in a toolpost and diameters are reduced, generally, by turning right-to-left rather than as for graver work.

The materials that clock repairers work with most are carbon steel and brass and we are likely to get away with just two tools for most basic work though a parting tool is also useful. Anyone making significant use of the lathe will have a variety of tool bits to work with, and would not necessarily use the tool sharpened in text-book style; but for the occasional use of the compound slide, two tool bits should suffice.

Tool bits are available from tool suppliers. For the clock repairer using a 6mm or 8mm lathe, a 5mm (³⁄₁₆in) tool bit should be a useful size. If you have two, four different tools could be ground, one at each end of each tool.

TURNING TOOL PROFILES

Most of the time we will be working with carbon steel such as Stubbs silver steel or KEA 108; or with brass, probably CZ 120 (or CZ 118 or 119); or standard brass. You may even make the occasional use of mild steel.

When a new tool bit is supplied it will need clearance angles ground onto some working faces, and sometimes a rake angle, or two rake angles. Here are some of the situations which should be considered before you grind a tool:

- Steep angles mean that faster feeds may be applied, but the tool will be weaker and will need grinding more frequently.
- Tools must cut, not rub.
- Angles that are too large can mean more work for the tool, so slower feeds will have to be applied.
- Steep angles can cause the tool to be pulled into the work with consequent loss of control and poor work finish.

- Sharp tools with a fine finish will mean better finishes on the job.
- Rounding the point on the tool leads to better work finish, but undercuts are more difficult to achieve.
- Finally, and perhaps the most important consideration of all, the tool must cut on the centre line of the work, not above.

The tool will be supplied with a clearance angle on the front; this must be a definite clearance, but need not be as much as is often seen.

When you grind, use safety glasses and make sure that you have safe equipment. Have some cold water available so that the tool bit can be quenched frequently; this will avoid a build-up of heat which could actually reduce the ability of the tool to cut.

A Tool for Cutting Steel

For the general turning of steel, grind the tool to the angles shown in Fig 46. For turning brass, note that there is no top rake on the tool – in fact if anything there is a negative rake. For parting steel, the other end of the tool bit could be ground as in Fig 48.

Setting the Tool

There is a variety of ways to hold lathe tools in the lathe – too many to list here. All ways involve securing the tool in some sort of tool holder, then adjusting the height of the tool to cut on, or slightly below, the centre line of the work. A well

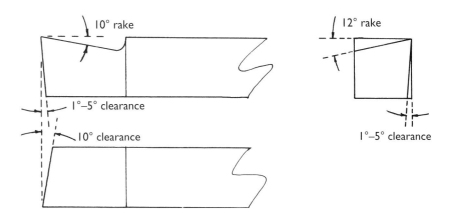

Fig 46 A lathe tool for turning steel.

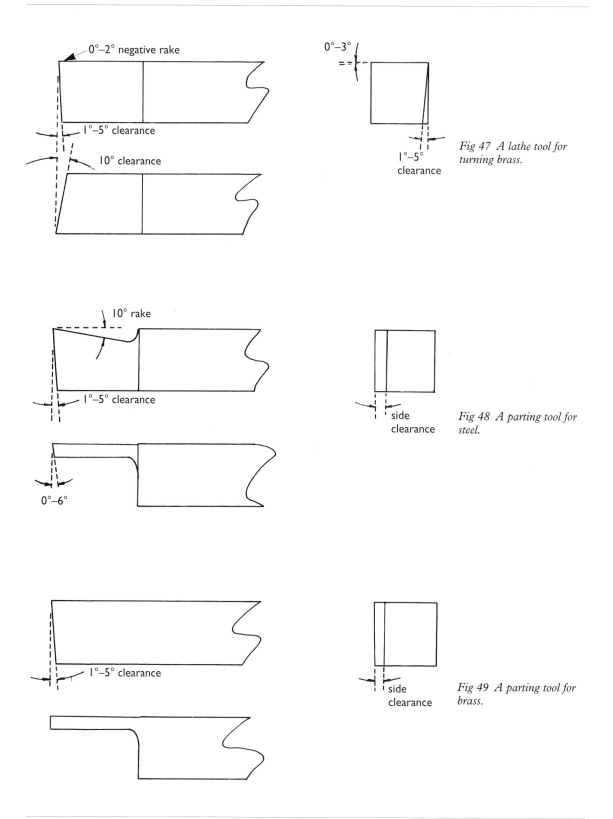

Fig 47 A lathe tool for turning brass.

Fig 48 A parting tool for steel.

Fig 49 A parting tool for brass.

Fig 50 Setting the tool to the centre line of the work.

tried and trusted method of checking the tool height is to rest a 150mm (6in) steel rule against the work, then bring the tool gently to the rule (with a thick piece of paper between the two to protect the tool). When the tool is the correct height, the rule will be upright or leaning towards you very slightly.

USING A CENTRE LATHE

If you propose using a centre lathe and you are inexperienced, in the interests of safety I would recommend that you have tuition through a short course of instruction. Only then should you consider using anything more powerful than a watchmaker's lathe for turning or gear cutting. The British Horological Institute run such short courses.

TURNING A NEW BALANCE STAFF

It is not intended that you should attempt to turn a balance staff immediately following the previous practice; however, having had experience of polishing pivots, repivoting and other lathe work, it is something worthwhile to be able to do. It can also save many platforms on carriage clocks from being condemned, thus contributing significantly to conservation and helping to keep the value of clocks.

A new balance is usually turned because the original has badly cut or broken pivots and a replacement is unavailable.

1. Confirm that the original staff was correct by assessing that the engagement between the roller and pallet fork was in order, that the balance was well clear of the pallet and escape cock or cocks, and that the arms of the balance were well clear of the boot. Basically it is a question of confirming that the original heights were correct.

2. Remove the collet, roller and balance from the original staff, following the recommendations in Chapter 12.

3. With a metric micrometer – although other units are still used, metric measurement is being used increasingly – measure the diameter of the hub of the old staff and select a piece of blue steel that is very slightly oversize.

4. Sketch the old staff and write down the dimensions, both diameters and heights; this is in case the old staff is lost. Making a staff without a pattern is possible, but it is much easier with a pattern or/and measurement to follow.

5. Mount the steel in the lathe with about 1½ times the staff length as overhang, turn it true, then down to the hub diameter.

6. Face off, if necessary, the end of the steel.

7. Holding your pattern (the old staff) with the balance seating against the end of the steel, mark on the steel the position of the end of the pivot.

If the balance pivot is broken, a judgement for length will have to be made; being slightly over-length here by about half a pivot is easily put right later.

8. Turn down the diameter to take the hole in the balance arms so that you have an interference fit. The seating must be slightly

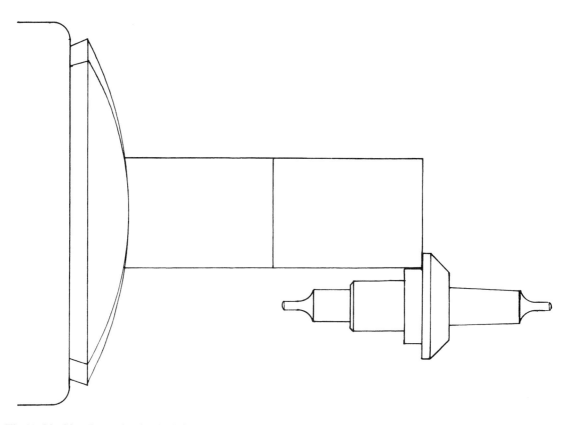

Fig 51 Marking the seating for the balance.

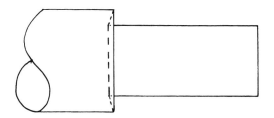

Fig 52 The finished seating showing the slight undercut.

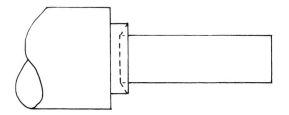

Fig 53 The collet diameter and steep undercut for riveting.

undercut to ensure that when fitted, the arms of the balance will lie flat against the seating.

9. Now we are going to turn the collet diameter and produce the means of riveting the balance staff. Start by marking on the steel with the point of your graver a distance, to the right of the seating, equal to the thickness of the balance arms plus sufficient for the rivet; about $\frac{5}{100}$mm should do. Turn the diameter for the collet producing a steep undercut for riveting the staff with a pointed graver. Leave the collet diameter oversize by about $\frac{1}{100}$mm for polishing later.

10. Offer up the pattern again and mark, with the point of your graver, the height of the collet diameter.

NOTE: Wherever possible, try to use the same reference point for measurement to avoid an accumulated error; usually the seating.

11. Turn down the diameter above the collet diameter to the pattern's dimension.

12. Offer up the pattern again and mark exactly where the pivot on the new staff is to start.

13. Remove the endpiece from the cock, and turn the pivot with a radiused corner so that it will almost enter the hole in the jewel. If the new pivot is very much too long, reduce the length with an Arkansas stone. At this stage the pivot should be, at the most, only one and a quarter to one and a half times the finished length.

NOTE: Now is the time to exercise extra care because we have reached a stage where the work achieved so far could so easily be ruined. Nevertheless, even if you do spoil your first attempt, take heart in the fact that few achieve a good, usable staff at the first try!

14. With an Arkansas stone, reduce the pivot diameter still further with a radiused corner

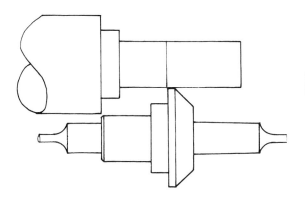

Fig 54 Marking the height of the collet diameter.

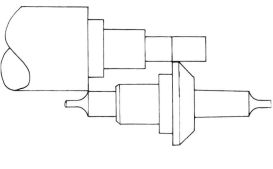

Fig 55 Mark where the new pivot is to start.

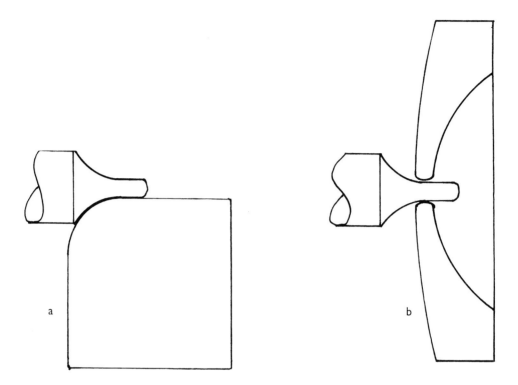

Fig 56 *Stoning the pivot (a), and checking for size (b). The side of the hole must meet the radius of the pivot.*

so that it is a snug fit into the hole in the jewel, and parallel. Don't enter the pivot into the hole in the usual direction, but from the convex side. The pivot must enter the hole all of the way until the radius at the bottom of the pivot meets the hole.

15. Burnish the pivot with the double-ended pivot file (burnishing end). Check that the diameter of the pivot would leave a 10° lean of the staff in the hole. If less, use the Arkansas stone and burnish again.

16. The most difficult part of the task is now over, and the risk of ruining the new staff is substantially reduced so you can polish other diameters now. There is no need to polish the diameter the balance fits onto. Polishing can be done with an Arkansas stone, followed by a burnisher or any other preferred method – for example, diamantine used with an iron or bell metal polisher.

We are now going to produce the hub and reduce the roller diameter as much as we dare before parting and finishing the hub, the roller diameter and the bottom pivot. For this, you should use either a pointed graver or a modified form of graver similar to that shown earlier in this chapter, but for left-handed use.

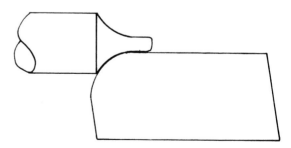

Fig 57 *Burnishing the pivot with the burnishing end of a double-ended pivot file.*

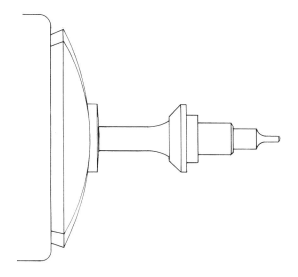

Fig 58 Turning the hub and roller diameter. Work from right to left so that the staff has support.

18. Next, part off, leaving a point on the end of the roller diameter which should leave the bottom part of the staff over-long by about a half-pivot length. The point is used as a visual check that the staff is true in the lathe before finishing the bottom half of the staff.

19. This part may present difficulties, but it is vital to a successful job. Hold the part-finished staff in the lathe again, but this time holding on the collet diameter. The staff must be held securely, dead true and giving access to your reference point on the seating. If the staff doesn't turn true, select another collet. If it still won't turn true, turn the staff round in the collet and try again.

20. Proceed to finish turning the hub by turning the backslope and producing a shoulder for the roller to bottom on.

21. Turn the roller diameter to a gentle taper such that the roller will fit onto the first two-thirds of it, but will have to be driven onto the remainder.

22. Next, mark the start of the bottom pivot (*see* Fig 60).

23. Remove the bottom endpiece from the bottom plate of the platform and produce a pivot as **steps 13** to **15**. Don't be tempted to use

17. Turn as much as is possible of the hub and roller diameter, but leave it oversize. This time work from right to left with your graver.

I will often turn the bottom pivot as well at this stage, although I would not recommend this if it is your first attempt at turning a staff.

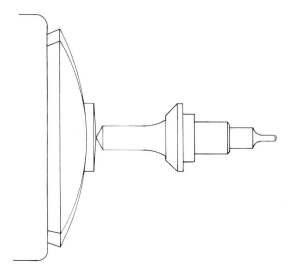

Fig 59 Part by turning to a point. This point is used for truing the staff to finish the bottom half.

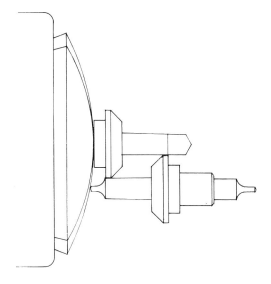

Fig 60 Marking the start of the bottom pivot.

the balance cock jewel to produce the bottom pivot, because the top and bottom jewels may be different sizes.

24. Polish the backslope and roller diameter, as **step 16**.

Again we come to a difficult stage, which is to check heights and finish the staff. With luck and good judgement we will simply have to remove the extra half-pivot length from both pivots, leaving them identical and well proportioned, and finally burnish their ends.

25. Remove the new balance staff from the collet, fit the balance and true it (*see* pages 129 and 131 for **steps 25** and **26**).
26. Fit the roller.
27. Replace the bottom endpiece, pallets and pallet cock.
28. Place the balance in the platform and carefully replace the balance cock but with the top endpiece left off. Don't tighten the balance cock screw as the staff should be a little too long and there is the risk of smashing the jewel hole or worse still, damaging the new staff.
29. Check in particular:
    • the clearance between the bottom of the roller and the plate;
    • the clearance between the guard pin and the bottom of the impulse pin;
    • that there is full engagement between the impulse pin and the notch;
    • the clearance between the balance arms and the pallet cock and screws;
    • that there will be clearance between the boot and the arms of the balance;
    • that there will be clearance between the stud and the arms of the balance;
    • the clearance between the top rim of the balance and the underside of the balance cock.

As the top endpiece and regulator are not in position yet, some will require judgement.

Hopefully you will find that you have sufficient scope to remove the extra half-pivot length you

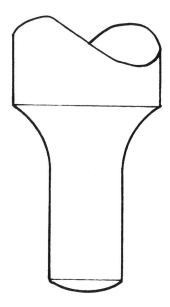

Fig 61 The end of the pivot should be rounded, but not semicircular.

have deliberately left on both pivots, and still leave safe heights.

30. Assuming that we have an ideal situation, remove the cock and balance. Holding the balance staff on the collet diameter, stone the end of the pivot to remove the half-pivot length, leaving it rounded but not semicircular.

    To ensure that the radiused part of the pivot cannot foul the jewel hole, a good check is to enter the pivot into the balance jewel without the endpiece, and see that the pivot shows above the convex side of the jewel and slightly above the plate.
31. Burnish the end of the pivot.
32. Replace the balance in the platform with the bottom endpiece in position, and replace the cock again but without the top endpiece. With the balance cock screwed down there should be up-and-down movement of the staff.
33. With the bottom pivot resting on the bottom cap jewel, stone the surplus from the top

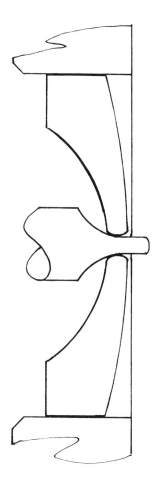

pivot until it is very nearly flush with the top of the top jewel hole.

34. Remove the balance staff from the platform again, and stone just a little more from the pivot; then before polishing, check again on the length. Ideally you will end up with the bottom pivot resting on the bottom cap jewel and the top pivot just lower than the top of the cock and about flush with the top of the jewel.

35. Proceed to burnish the end of the pivot holding on the large roller diameter without reducing the length. In general, burnishing smooths metal over, it doesn't remove it; however, the pivot is so fine that it *is* possible to remove metal by too much burnishing. When shaping the end of the pivot, try to obtain the same shape on the ends of both pivots. This is less vital for a timepiece that isn't portable, because then the balance always rests on the one pivot so the need for similar frictional losses on both pivots doesn't arise.

36. Once you are satisfied that all is well, poise the balance and continue to reassemble as for fitting a new staff. A broken balance staff is often, though not always, accompanied by cracked jewel holes or endstones. Select and replace broken jewels before turning a new balance staff. It is preferred that both pivots are of the same diameter.

Fig 62 Check that the rounded root of the pivot cannot foul the jewel.

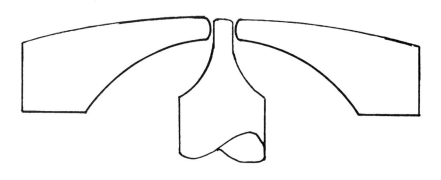

Fig 63 The radiused part of the pivot clearing the hole in the jewel, with safe clearance allowing for endshake.

6 REPAIRING WORN HOLES AND PIVOTS

For the efficient transmission of power in a clock, it is important that meshing wheels and pinions continue to operate at, or close to, their correct distance of centres. Pivot holes do wear and as a general rule, when a hole becomes worn by about one fifth of its diameter, a repair is necessary. This usually involves opening the hole to accept a bush which will return the pivot to its original position.

Opening a hole is normally done making use of a five-sided cutting broach, working from the inside of the clock plate and turning the broach in one direction only. Finishing the hole to accept the pivot can be effected by using a smoothing broach; this compresses the brass around the hole, giving a harder bearing and smoothing the hole, thus reducing friction.

IDENTIFYING A WORN HOLE

If necessary take some of the power off the great wheel, then reverse it by holding it between index finger and thumb and turning it backwards, a worn hole will often then show up by a pivot jumping across it. The likely direction of wear is illustrated in Fig 65: 'A' represents the driving wheel and turns in a clockwise direction; 'B' represents the driven pinion; 'C' is the usual direction of wear. If the driving wheel 'A' turns in an anti-clockwise direction, the wear will be on the other side of the line joining the centres, and again, just off the right-angle. Pinions are frequently nearer one plate than the other, and it is usually the hole at the pinion end of a wheel that wears.

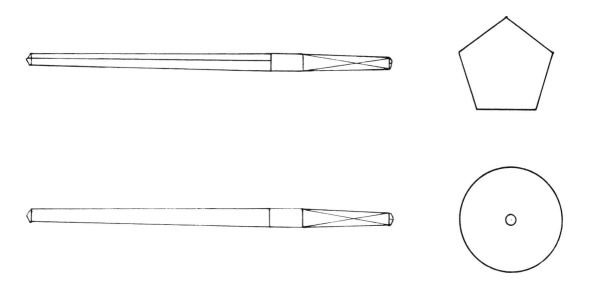

Fig 64 A five-sided cutting broach and a smoothing broach.

*Fig 65 A wheel 'A' driving
a pinion 'B'. The direction
of wear is most often 'C'.*

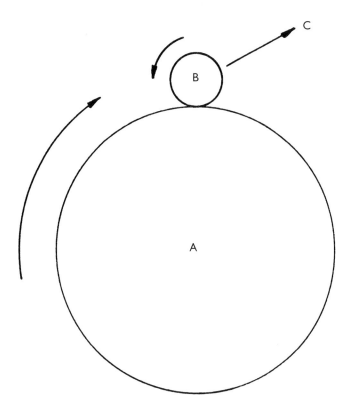

Bushes are sold in different ways, and it is possible to buy assortments for particular types of clock and packets of one particular size. They will have at least three dimensions: a hole size, a diameter and a height. Some have a taper similar to the taper of a broach, others have no outside taper, and there is no particular preference for one over the other. For the beginner to clock repairing, I would recommend a small assortment of bushes selected from a catalogue to suit the type of clock being worked on, then to extend the range as different clocks are encountered.

Having identified which holes are to be bushed in all trains, and having stripped the clock, proceed as follows, remembering the secondary objective of leaving minimal evidence that the clock has been worked on. Try to work within the limits of the original countersink.

1. Select a bush of appropriate diameter, which should be no greater than the diameter of the countersink, and of a height equal to the thickness of the plate. If the height is greater than the plate thickness, reduce the height of the bush in the lathe.

2. If possible, mark the countersink on the opposite side to the wear, a distance equal to the wear.

3. With a rat-tail file, file the good side of the hole to your mark; this is known as 'drawing the hole'. If it is not practical to mark and file, draw the hole by broaching, putting pressure on one side of the broach. As an alternative, for a small hole, a piercing saw blade held in a pinchuck could be used to draw the hole.

4. Now put a suitable broach into a pinchuck and broach the hole from the inside of the plate, turning the broach in one direction only, keeping it at right-angles to the plate. Avoid putting forward pressure on it. If, in the process of doing this, it becomes five-sided, as some do, bring the hole back to round with a rat-tail file, then carry on broaching.

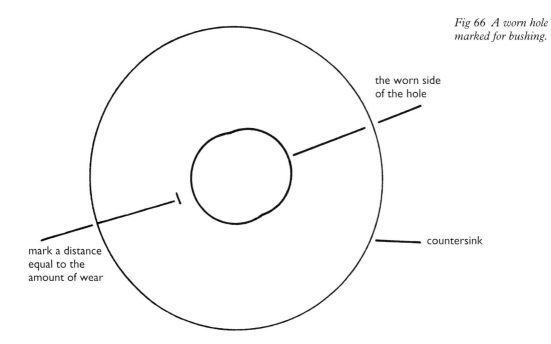

Fig 66 A worn hole marked for bushing.

the worn side of the hole

mark a distance equal to the amount of wear

countersink

By now the plate will have a burr on the inside which can be ignored for the moment. Continue broaching until the hole is round and the pre-selected bush, which may be parallel or tapered, just starts to enter the broached hole from the inside of the plate. The bush is fitted from the inside of the plate to reduce the possibility of it falling out, and thereby causing more damage to the clock.

5. Remove the bush and deburr the inside of the hole with a roller sinker or countersink tool. A drill can be used for deburring if used carefully.
6. From a staking outfit, select a punch with a flat end which is larger in diameter than the bush.
7. Place the bush back into the hole, pushing on it with the back of the tweezers until it is just held in the plate.

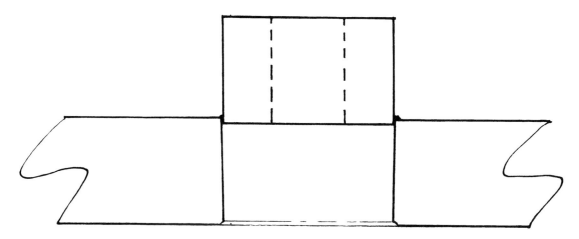

Fig 67 The bush just entering the broached hole.

Fig 68 Centering the bush over a suitable stake.

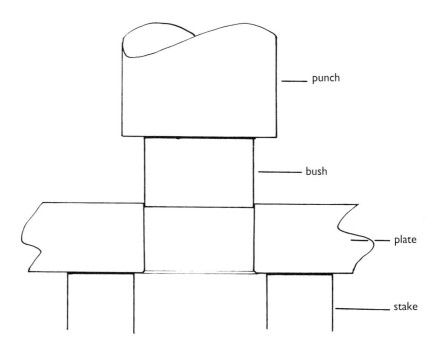

8. Centre the bush over a hole in the stake which is slightly larger than the diameter of the bush.
9. Put the flat-ended punch in the stake and drive the bush in until it is flush with the plate, which will happen when the punch contacts the plate. As this occurs, a dull solid sound should be heard. Some repairers rivet bushes in a plate, but I very rarely find this necessary.
10. Using a cutting broach held upright and entered from the inside of the plate, open the hole until the pivot fits. A good fit is indicated by a lean of 5° of the wheel in all directions, with the shoulder of the pivot resting against the plate. The final little bit of broaching may be done with a lubricated polishing broach, but it is not essential.
11. Finally, deburr the inside of the hole and countersink the outside of it so that the countersink of the bush blends in with the original countersink. The bush should now be virtually invisible from the outside of the plate.

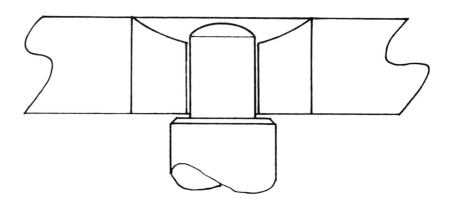

Fig 69 The finished bush.

When countersinking, retain as much height as is reasonably possible around the pivot, yet the pivot must come through the hole even when allowing for endshake.

If the bush you have picked out is much too high for the plate, reduce its length in the lathe before fitting. If a selected bush is only slightly high for the plate, its length can be reduced with the countersink tool after fitting. Avoid filing a bush after it has been fitted, however, as the file tends to leave unsightly marks on the plate.

Not all pivot holes are countersunk – for example, the holes for the barrel of long-case clocks – and in such instances the previous instructions should be modified. For example, the broached hole should be deburred on both sides of the plate before a bush of the same thickness as the plate is driven in.

There will be occasions when a barrel hole or barrel cover hole may need to be bushed. These circumstances will probably require the barrel to be held in the soft jaws of the self-centring chuck, and the hole to be bored out to take a bush.

REPAIRING DAMAGED PIVOTS

Brass is porous, and it will harbour small particles of hard material which, when mixed with oil, as happens when oiling pivot holes, will cut a steel pivot. Here we will look at different ways of repairing damaged pivots, and how to renew a badly cut pivot.

The arbors, pinions and pivots of better quality clocks are made of carbon steel, hardened and tempered to dark blue. Some will be made of carbon steel but not heat-treated, while others are made of mild steel where only the outer skin can be hardened (case hardening).

How to Recognize a Good Pivot

There are three types of pivot used in clocks: **straight**, generally used for train wheels; **cone**, reserved for the balance staff of less expensive clocks; and **conical**, mostly used on the balance staff of better quality clocks but may be found on the escape wheel and pallets as well.

The straight pivot should be slightly longer than the hole in the plate, including the endshake; it should be straight with parallel polished sides, and it should have a square shoulder and a rounded but not semicircular end.

The cone pivot should be polished with a 45° included angle meeting at a virtual point. The point should have a slight radius of about $\frac{2}{100}$mm.

The conical pivot is polished with a rounded shoulder, has straight sides and a rounded end.

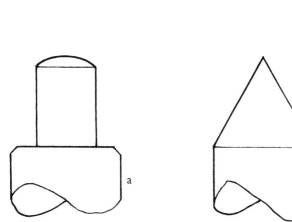
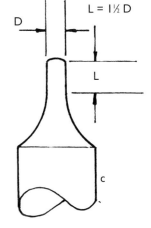

Fig 70 (a) Straight pivot. (b) Cone-shaped pivot. (c) Conical pivot.

Polishing a Badly Cut Straight Pivot

1. Mount the wheel in a lathe with the minimum of overhang, taking care to ensure that the uncut part of the damaged pivot is turning dead true. This is important.
2. Turn the pivot down by the minimum necessary to make it parallel, but leaving a square shoulder; avoid turning the shoulder back, or the endshake of the wheel may become unacceptably large.
3. Rub two sides of a triangular Arkansas stone on emery cloth until they meet at a sharp angle, then clean the stone and lubricate it with medium clock oil.
4. Clean the turned pivot, then stone it to remove any turning marks until a uniform matt grey finish is achieved. Apply the stone with a fairly high speed set on the lathe and firm pressure on the stone. Keep it moving forwards and backwards, avoiding any rounding where the pivot meets the shoulder, but do break the sharp corner where the parallel part of the pivot meets the rounded, but not semicircular end.
5. Clean the pivot with degreasing agent and paper towel ready for burnishing.
6. Put a light grain across a flat burnisher with a No 2 buff stick, wipe off the dust and smear the burnisher with a medium grade, clean clock oil.
7. Have the lathe turning at a fairly rapid rate, then bring the burnisher to the pivot applying a fairly heavy pressure while moving the burnisher quickly forwards and backwards, and occasionally introducing a circular motion which helps to prevent lines forming on the pivot. As the polish improves, maintain the speed of the lathe and burnisher but ease off on the pressure. Take care to keep the burnisher at right-angles to the arbor, otherwise its side might act like a turning tool and cut the shoulder.

This technique can produce good results. Alternatively a flat clean piece of wood could be charged with Autosol (available from motor accessory shops) and used in a similar fashion.

Another traditional method of polishing is to charge iron or bell metal with diamantine-and-oil mix, and to polish as above.

NOTE: It will probably be necessary to bush a pivot hole after turning and polishing a pivot.

Polishing Lightly Cut Pivots

A lightly cut pivot is polished as described above from **step 4**.

Polishing Cone-Shaped Pivots

1. Mark the rim of the balance with a light 'prick' from a scribe at the point where the balance spring terminates. Also note how far the collet is driven onto the balance staff. This is to help reassembly so that the collet is positioned correctly.
2. Drive the balance staff from the collet (Fig 71).
3. Mount the balance in the lathe with minimum overhang.
4. Stone the damaged pivot with an Arkansas stone until a 45° angle with a sharp point is achieved (Fig 72). The point should be so sharp that if you draw your thumb lightly across it with an engaging action, it can be felt to catch your thumb print. (It is not intended to tear the skin.)
5. Now burnish the pivot after cleaning (Fig 73). Once the desired finish is achieved, roll the burnisher lightly round the end of the pivot to give it a radius of, say, $\frac{2}{100}$mm.
6. Remove the balance from the lathe and polish the other pivot in a similar fashion. It is usual to polish both pivots so that frictional losses on balance pivots are the same top and bottom.
7. Replace the collet on the staff to the original position, supporting the staff on an iron block with a hole and driving the collet on with a flat hollow punch (Fig 75).
8. After putting the balance back into the clock, check that the clock is in beat.

Polishing Conical Pivots

Polishing conical pivots is not normally recommended as this may involve changing the balance jewels in the clock to obtain a good fit. Instead, consider fitting a new balance staff.

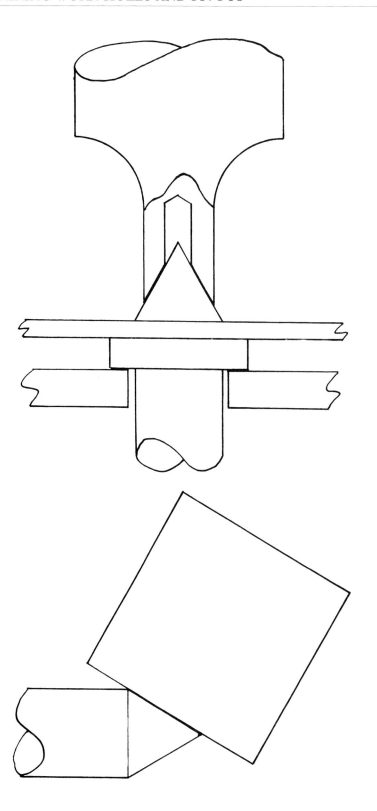

Fig 71 Driving the staff from the collet.

Fig 72 Stoning a cone-shaped pivot.

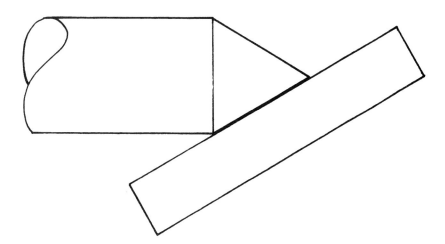

Fig 73 Burnishing the pivot.

A distinction needs to be made between polishing and burnishing: polishing involves the removal of metal; burnishing on the other hand is smoothing the metal over, imparting a polished-looking, hard skin. Conical balance pivots may be burnished, which involves removal of the collet and roller (*see* pages 127 for this). First, however, the double-ended pivot file and burnisher needs explaining: as its name suggests, this tool is indeed double-ended, one being for burnishing, the other for filing. There is a Swiss type and an English type.

A close inspection of an English right-handed tool will reveal a section through the tool that is like a rectangle pushed out of shape; this is so the user can get up close to a shoulder when working right-handed. Even closer inspection will reveal that while one side of the burnisher has sharp corners, the other side has one corner which is 'rounded smooth' at its outer edge, but rounded with a file edge close to the handle. The file end has one side with sharp corners, whilst the other side is rounded and polished on its outer edge, and rounded with a file edge close to the handle.

1. Mount the balance in the lathe holding on the roller diameter with the minimum of overhang necessary.

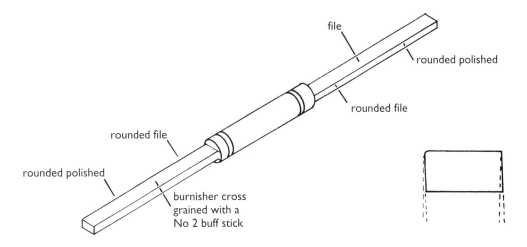

Fig 74 A double-ended pivot file (left), and a section through a double-ended pivot file and burnisher (right).

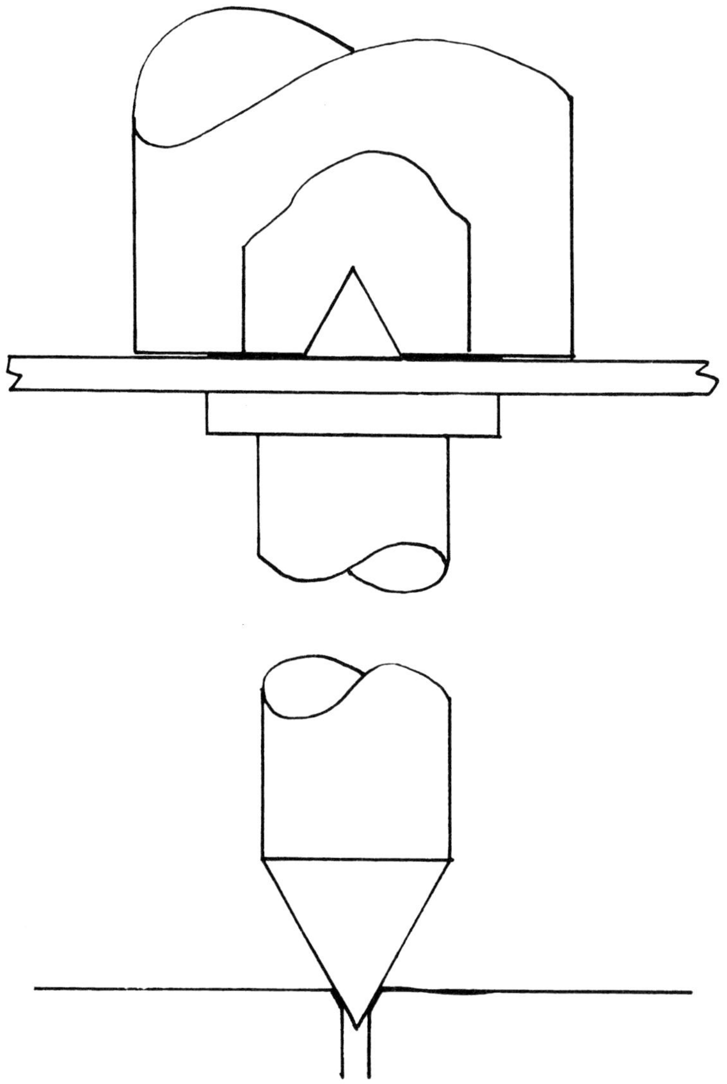

Fig 75 Driving the collet back onto the staff.

REPIVOTING

2. Use the lubricated pivot burnisher under-hand, with its rounded plain edge against the shoulder of the pivot, in much the same fashion as for polishing flat pivots.

3. Burnish the end of the conical pivot by rolling the burnisher round the end of the pivot. The pivot is so small in diameter that there is a risk of removing some metal.

4. Now hold the staff on the collet diameter and polish the other pivot. It is desirable to keep the ends of both pivots to the same shape.

There is no hard-and-fast rule about just when a cut pivot needs renewing altogether, although there are two significant factors to bear in mind: first, when a pivot is reduced in diameter by one half, it becomes one eighth of its original strength; and second, where the power is stronger, for example on an intermediate wheel, the scope for reducing the pivot diameter is less due to the possibility of it breaking.

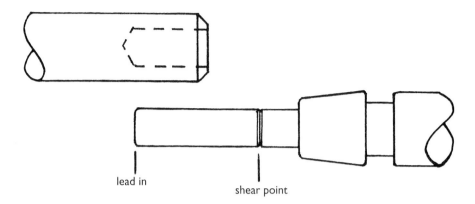

lead in

shear point

Fig 76 The drilled hole, plug and shear point.

1. Start by testing the arbor to be repivoted for hardness. A correctly heat-treated screwdriver is hardened and tempered to blue, so if a good screwdriver will mark the arbor then this is likely to be turned and drilled with comparative ease. If the screwdriver will not mark it, draw the temper of its end until the graver will pick up a centre; often tempering to light blue or just beyond is sufficient. Keep the heat away from the brass wheel to avoid softening it.

2. Mount the wheel to be repivoted in the lathe with the minimum of overhang necessary, and with the root of the pivot turning dead true; this is important if the new pivot is to be concentric with the wheel and pinion (assuming that the original was concentric).

3. Select a high-speed twist drill, or make a spade drill the same diameter as the intended pivot, and secure it in a pinchuck or pinvice. Check that the drill is sharp.

4. Pick up a centre with a graver such that the web of the drill will enter the centre, but not the whole diameter of the drill.

5. Drill a hole to a depth of one-and-a-half to two times the length of the original pivot, keeping the pinchuck parallel to the bed of the lathe. If the arbor resists the drill, obtain a tungsten carbide drill and try with that. I always keep a range of such drills between 0.5mm and 1.5mm and have always enjoyed success with them. Avoid using cutting lubricants because the plug is to be seized in the drilled hole later.

6. Now file a plug for the drilled hole using a filing block and No 4 double cut file. If the arbor was hardened and tempered, make the plug from blue steel. If the arbor was left unhardened, make the plug from carbon steel in its soft state.

NOTE: File the plug so that it is parallel with a diameter that you feel will be a tight interference fit in the drilled hole. A slight lead-in on the plug is desirable. Be careful not to stretch the end of the hole when trying the plug for fit.

7. With top cutters or a file, put a light cut around the plug in a position which is just a little longer than the depth of the hole plus the length of the pivot. This is the intended shear point of the plug. Some repairers make a plug with a $\frac{2}{100}$ to $\frac{3}{100}$ taper and drive it in, but this has the disadvantage of splitting some arbors, and it can work out when the pivot is being worked on.

8. File four flats on the part that is being held in the pinchuck to avoid it turning in the pinchuck.

9. Turn the lathe by hand, at the same time pushing the plug to the bottom of the drilled hole. Keep turning the lathe by hand until you can feel the plug getting tighter. Eventually, after perhaps four turns of the lathe, the plug should seize in the hole and break off at the reduced diameter. If it doesn't, draw it out again and dip the end in dry oilstone dust and repeat the operation. My experience is that plugs always shear after this treatment.

10. When the plug shears, the strain may bend the pivot slightly. If this happens, bend it straight, then if necessary, turn the pivot true and polish as previously explained. Bending is best effected with bran lined pliers which is less likely to mark the steel.

Ideally the new pivot will be just larger than the original hole so that it can be lightly broached to remove any hard particles embedded in the brass; the hole could then be finished with a round smoothing broach. It should not be necessary to bush the hole.

7 STRIKING CLOCKS

Basically there are two types of striking clock, the rack-type and the locking plate. The former is much easier to set up and is largely self-correcting if the time and strike are put out of step. The latter on the other hand frequently requires certain wheels to be correctly associated by moving one wheel in relation to another after preliminary assembly; it is not self-correcting. Instead, the hands need to be adjusted to catch up with the strike, or the strike train will need to be operated independently to catch up with the time shown on the dial.

With both rack striking and locking plate, in addition to a time train there is a second train, usually to the left of the time train when viewed from the dial side, called a strike train. In a rack strike the train usually consists of a barrel, an intermediate wheel, a hammer-lifting wheel, a gathering-pallet wheel, a warning wheel and a fly to control the speed of the train and to dissipate the energy built up when the train comes to a halt at the end of striking. Other significant parts of a rack strike include a lifting piece, a rack hook, gathering pallet, warning detent and a locking detent.

COMMON PRINCIPLES

Although striking clocks do vary slightly in operation, certain principles remain common to all. Each time the strike train is operated, there must be a short run before the hammer starts to lift to cope with the inertia of starting all the wheels in the train from stationary, and to cope with the inertia of lifting the hammer. A further slight run of the train after the last blow has struck is necessary to ensure that the hammer does fall after the last blow.

To prevent uncertain unlocking of the strike train, a warning is given which involves a half turn of the warning wheel prior to the eventual release of the train for striking. During this period of warning, the hammer must not be 'on the rise' (a term used to indicate that the hammer is being lifted by the hammer-lifting wheel).

To understand striking clocks it is recommended that you have a Smiths mass-produced striking clock to look at and work on. Again, the Smiths has been chosen because of its popularity and availability; also because what is learned is transferable to other striking clocks.

A RACK-TYPE STRIKING

To understand more clearly how the striking train operates, make use of Figs 77a and 77b. For clarity, the hour wheel is removed in Fig 77b.

Riveted to the cannon pinion are two lifting cams, one longer than the other. When the long lifting cam triggers the strike train through the lifting piece, it allows the rack to fall with the possibility of more than one blow being struck. When the short cam triggers the striking train, the rack will not fall, and consequently only one blow is struck.

The train is locked due to a projection on the rack hook called the locking detent, which engages with the warning-wheel pin. At the same time another projection on the rack hook lies in the gap in the gathering pallet, and the hammer arbor – which can't be seen – is clear of the hammer-lifting wheel by about 4mm (³⁄₁₆in). These relative positions are important.

As the cannon pinion rotates, the high lifting cam lifts the lifting piece which in turn lifts the

Fig 77a A rack-type, striking-clock movement.

Fig 77b The clock with the hour wheel removed for clarity.

rack hook. They continue to lift until the locking detent releases the warning wheel. But the warning wheel, having made half a turn, is then arrested by the warning detent on the lifting piece. During this operation, called 'the warning', the hammer must not be on the rise. This is very important.

The lifting piece and rack hook continue to rise until the rack hook frees the rack which drops until the rack tail contacts the snail. The snail governs the number of teeth released and consequently the number of hammer blows.

The lifting piece and rack hook have a further slight rise until the lifting piece falls off the lifting cam. Both rack hook and lifting piece fall, releasing the warning wheel allowing the strike train to operate. Still the hammer arbor must not lift until

the striking train has sufficient momentum to cope with the extra effort of lifting the hammer. Usually this is about half to one-and-a-half turns of the warning wheel.

The clock now continues to strike until all the teeth of the rack have been gathered up. So far the train hasn't relocked because the rack teeth are too shallow to allow the rack hook to fall into the gap in the gathering pallet. When all of the teeth are gathered up by the gathering pallet pin, the rack hook can drop into the gap in the gathering pallet allowing the locking detent to come into the path of the warning pin.

At the half hour, the low lifter operates the striking train but will not lift sufficiently high for the rack to fall; consequently only one blow will be struck.

DISMANTLING A STRIKING CLOCK

1. Remove the hands: they will be screwed or pinned.
2. Remove the pendulum.
3. Remove the gong; there is usually one nut under the case.
4. Remove the case screws while protecting the glass, and lift the movement out of the case.
5. Remove the suspension spring and suspension rod immediately after removing the movement from the case; this is to protect the suspension spring.
6. Examine the movement for wear, and establish the cause of failure.

As you gain experience you will want to tackle different movements. If, however, the movement is different from those you are familiar with, study it carefully to ensure you know how it functions, and *do not* strip a clock until you can explain the function of each part.

7. Remove the power from both trains. Use a well fitting letting-down key and exercise great care. Avoid over-compression of the click spring, and be careful not to damage the ratchet wheel teeth.
8. Remove the pallet cock screws, pallet cock and pallets.
9. Remove the hammer.
10. Remove the rack.
11. Remove the pin and washer securing the minute wheel.
12. Remove the hour and minute wheels.
13. Remove the rack hook.
14. Remove the lifting piece (if it will pass the gathering pallet).
15. Remove both ratchet wheel bridge screws, bridges and ratchet wheels. Discreetly mark 'T' and 'S' to identify time and strike positions. Some parts on quality clocks are marked with one centre pop for time, two for strike.
16. If screwed, remove click and click springs. Also mark 'T' and 'S'.
17. Remove the four pillar nuts from the back plate. Always try to leave the movement retaining brackets undisturbed.
18. Remove the back plate, taking care not to disturb the trains. Some parts need identification marks.
19. On the strike side remove: fly, warning wheel, hammer arbor and hammer-lifting wheel. These need no identification marks.
20. Remove and mark the strike intermediate wheel 'S'. (Actually, on the Smiths this wheel is much smaller than the time intermediate wheel, but on other strikes they could be very similar.)
21. Remove the strike barrel and mark 'S' on the barrel, cover and arbor. Identify the cut-out in the barrel cover with the wall of the barrel. (On some clocks the barrels will not come out until the cannon pinion is driven off or until the centre wheel clutch arrangement is dismantled.)
22. On the time side, remove the escape and fourth wheel.
23. Remove and mark 'T' on the time intermediate wheel.
24. If possible remove the time barrel and mark.
25. Remove the gathering pallet with support under the plate, a suitable punch and mallet.
26. Still holding over the bench, remove the cannon pinion in a manner similar to that described in step 11, 'Dismantling', Chapter 3, placing a hand on the square to prevent the centre arbor digging into the mallet. Alternative ways of dealing with the cannon pinion are to dismantle the centre wheel as in **step 21** above, or to leave it *in situ*. If you can see that no repairs are necessary to the centre pivot or hole, and if you can clean, dry and lubricate without removing the cannon pinion, then leave it. This is a recommended short-cut.
27. Remove the barrel covers, arbors and mainsprings as described in 'Dismantling', Chapter 3.

GENERAL REPAIRS

If necessary, complete any repairs before cleaning. Common faults include worn holes, torn mainspring ends, damaged barrel teeth, damaged pinion leaves on the intermediate wheels, and cut and bent pivots.

CLEANING

Parts are cleaned in the usual way (*see* 'Clock Cleaning', Chapter 3). Peg all pivot holes after cleaning.

REASSEMBLY

1. Replace the mainsprings into their barrels, preferably with a mainspring winder although replacement by hand is possible (*see* 'Reassembly', Chapter 3).
2. Lubricate the mainspring with heavy oil. Use sufficient to give the whole mainspring a thin film.
3. Replace both barrel arbors, ensuring safe hooking.
4. Replace both barrel covers, associating cut-outs with the marked barrel walls.
5. Check endshakes on both barrel arbors.
6. Lubricate the diameters on the arbors which contact both barrel and plate.
7. Replace both barrels and trains, including the hammer arbor. Reassemble the centre wheel if dismantled. Lubricate the inside of the centre-wheel pinion where it fits onto the centre-wheel arbor.
8. Replace the top plate and four pillar nuts, making sure that the lifting-piece spring passes through the correct hole in the front plate.
9. Holding the plates horizontally, check the freedom and endshakes by lifting each wheel in turn and checking that each falls under its own weight.
10. Lubricate the centre-wheel front pivot with heavy oil.
11. Replace the cannon pinion if previously removed. Temporarily replace the hour wheel and minute hand to ensure there will be endshake in the hour wheel when the minute hand is secured. Having checked, remove the minute hand and hour wheel again.
12. Lubricate both ratchet wheels and ratchet wheel bridges with heavy oil (H), and replace them. If removed, replace clicks and click springs. Lubricate points of friction with heavy oil. Check that the action of the click springs is safe. This means that when the clicks are fully engaged with the ratchet wheels, there is still some pressure of the click spring on the click.
13. Lubricate the striking train as follows:
    Top and bottom intermediate wheels (H).
    Top and bottom hammer-lifting wheel (H).
    Top and bottom gathering-pallet wheel (H).
    Top and bottom warning wheel, medium oil (M).
    Top and bottom fly (M).
    Hammer arbor pivots, stop and lifter (H).
14. Lubricate the time train as follows:
    Top and bottom intermediate (H).
    Top centre wheel (H).
    Top and bottom third wheel (M).
    Top and bottom escape wheel (M).
15. Put the hammer arbor under tension with its spring lightly loaded. Only a light load is necessary as the hammer is planted horizontally and is already assisted by gravity.
16. Put a little power on the strike train by winding, and check that there are two-and-a-half to three turns of the warning wheel from the arbor falling off one lifter to being picked up by the next. Adjust if necessary by bending the stop.
17. Replace the lifting piece, securing it with its pin. Lubricate (M).
18. Replace the rack hook and pin. Lubricate (M).
19. The gathering pallet, in addition to gathering the teeth in the rack, is indirectly responsible for locking the strike train (and directly on some clocks). Fortunately it is secured to a round extended pivot on the Smiths strike and so it can be positioned just where we want. Ideally, to accommodate slight errors in manufacture and wear, a warning wheel should advance a further quarter turn after

the last hammer has struck, and then lock. This quarter of a turn is a minimum in our clock, and the run-on after the last hammer has fallen may be up to very nearly one-and-a-quarter turns.

To position the gathering pallet correctly, hold the rack hook up out of the way and control the strike train by intercepting the fly to establish the precise moment the hammer arbor falls. Immediately the arbor falls, stop the train. Now establish where the pin in the warning wheel is. If by chance it is in the ideal position of a quarter turn before its normal stopped position (this being 180° away from the warning detent), then simply allow the train to advance so that the pin in the warning wheel is at its normal stopped position; then put on the gathering pallet to allow locking to take place.

In the event that the hammer falls when the pin in the warning wheel is less than a quarter turn from its locked position, allow the warning wheel a further turn to come to its stopped position.

Should the pin be after the locking detent but before the last quarter turn, simply allow the warning pin to advance to the locking detent. On different clocks you will have a minimum of a quarter turn on the warning wheel when the last hammer falls, but up to nearly one-and-a-quarter turns. Long-case clocks and some others, however, will not accommodate this wide variation,

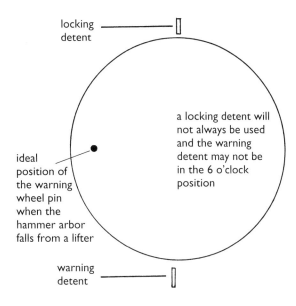

Fig 79 *The ideal position of the warning pin in relation to the locking detent when the hammer arbor falls from a lifter.*

and furthermore the gathering pallet may fit onto a square. In these cases, the power must be removed and the wheel positions changed with respect to one another so that the hammer arbor drops off a lifter with only a quarter turn of the warning wheel to its locked position, 180° from the warning detent.

20. After securing the gathering pallet with a hollow punch and support punch, you should check all of the lifters to ensure that the hammer arbor falls safely before the warning pin reaches the locking detent.

21. Now check that there is half a wheel warning and that during the warning, the hammer is not on the rise.

22. Lubricate, with medium oil, the minute wheel post and the post for the rack.

23. Replace the minute wheel, hour wheel and rack, but do not pin yet.

24. Position the minute hand on its square and turn it until the high lifter releases the rack from the rack hook. Hold the scape wheel stationary during this operation.

25. Still holding the time train stationary, position the hour wheel and snail so that the tail

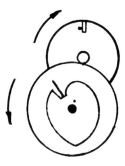

Fig 78 *The position of the gathering pallet relative to the rack hook when the train is locked.*

of the rack rests in the centre of a flat for any hour between two and eleven. Adjust if necessary by disengaging the hour wheel from the minute pinion, or to get a fine adjustment, leave the hour wheel engaged with the minute pinion, but disengage the minute wheel from the cannon pinion and turn the minute wheel. This has the effect of turning the hour wheel by part of a turn due to the gear ratio. Test particularly that the rack falls safely on the 11, 12, and 1 o'clock, spots – although not necessarily centrally on these.

26. Replace the washer on the minute post, and pin. The washer is to prevent the hour wheel dropping forward during casing and disassociating with respect to the cannon pinion. Check the endshake of the minute wheel.

27. Replace the rack, pin it, and check endshake.

28. Lubricate the friction points between lifting piece, rack hook and gathering pallet with medium oil. Also oil the rack teeth and tail.

29. Replace the pallets, pallet cock and screws.

30. Put power on the time train and adjust the pallet cock to give 1° of drop. Keep the pallet cock parallel.

31. Lubricate both pallets and pallet pivots with medium oil.

32. Replace the suspension rod and suspension spring.

33. Lubricate, with medium oil, the position where the crutch contacts the suspension rod. There must be a small clearance here.

34. Replace the hammer and pin.

35. Wind both trains fully.

36. Clean and polish the clock case and glass.

37. Replace the movement in the case and secure. To avoid screws coming through the front of the case, check the length of new screws; probably 9mm (⅜in) No. 4 wood screws would suit most mass-produced strikes.

38. Replace the gong.

39. If necessary, turn the hand in relation to the collet to strike exactly at the hour. This is done by passing the collet over the tang of a broach held in a vice, and turning the hand on the collet. To avoid breaking the hand, push it round with the top cutters applied to its root near the collet. Remember to remove the broach from the vice immediately after use, in the interests of safety.

40. Replace the hour and minute hands, checking clearances between dial and glass.

41. Replace the pendulum.

42. Put the clock on a level surface, put it 'in beat' and to time. Test over seven days.

LOCKING PLATE STRIKING

For simplicity, a long-case locking plate striking mechanism has been chosen to explain how

Fig 80 The counting lever on completion of striking an hour.

locking plate striking operates. The main principles of operation remain the same as rack striking, although the detail is different. Still there is half a wheel warning; still the hammer must not be on the rise during the warning.

The striking train has a great wheel which doubles as a hammer-lifting wheel. In the instance of the clock being described, there are thirteen lifting pins, a hoop wheel which acts as a locking wheel, a warning wheel and a fly.

On the front plate is a lifting piece, operated by a pin in the minute wheel which turns once an hour, and a warning detent. Other parts are ignored for the moment as they do not contribute to understanding the operation of the striking mechanism.

On the back plate placed on a square, which is an extension of the great wheel arbor, is a pinion which drives the locking plate. At the end of striking, a counting lever, sometimes called a locking lever, drops into a gap in the locking plate allowing the locking detent to arrest the hoop wheel. For a clock that strikes hours only, there must be twelve revolutions of the hoop wheel for 12 o'clock, one for 1 o'clock, two for 2 o'clock, and so on; So for a full twelve hours the hoop wheel must turn $12+11+10+9+8+7+6+5+4+3+2+1$ times – that is, seventy-eight turns. As the locking plate turns once each twelve hours, it can be thought of as having seventy-eight divisions.

When the striking train is correctly set up and the last blow for any particular hour is struck, the locking detent should be in the hoop wheel almost ready to lock. The train should run on for about half a turn of the warning wheel leaving the pin in the warning wheel half a turn away from the warning detent. When the clock next strikes, the hoop wheel is unlocked and the warning wheel advances half a turn for the warning. When the lifting piece falls from the pin in the minute wheel, the train is released and after a brief turn of the great wheel, the hammer is lifted.

To strike more than one blow, the counting lever is held in an elevated position, preventing the locking detent from arresting the hoop wheel. If the clock is a half-hour strike, there are wider gaps in the locking plate to allow for locking at the half-hour; in this case the locking plate can be thought of as having $78+12$ divisions – that is, ninety.

To help in setting up the locking plate and its driving pinion, the square on the great wheel is often marked so that the pinion is correctly associated with the square, and a leaf of the pinion has a dot or chamfer to associate with the wheel riveted to the locking plate. It is important to get the association right, or mislocking is likely to occur. When marks are not present, a search will have to be made to find the best association. You don't have to try every tooth on the locking plate into a leaf of the driving pinion; usually a search over seven or eight consecutive teeth will leave the locking lever safely in the centre of a gap for

Fig 81 The locked position for the train showing the hammer clear of the lifting pins, the hoop wheel locked and the warning wheel pin 180° away from the warning detent.

an hour strike, and engaging safely for the hour and half-hour of a half-hour strike.

When setting up a locking plate train, if the wheels are not correctly associated, it will be necessary to open the plates slightly and disengage a pinion from a wheel and move it round until the correct tooth space ties up with the correct pinion leaf. If the clock is spring-driven, it will be essential for the clock's safety and your safety to remove all power from all trains first.

It is easy inadvertently to bend a pivot when moving one wheel in relation to another. Avoid leaning a wheel by more than its natural lean of about 5° from vertical or the pivot will be strained, with the probability of bending or breaking it.

Procedure for Opening the Plates

Let us assume for the moment that the plates to be opened are secured with screws or nuts. After removing all power from all trains, start by slackening the two lower fixings by just a couple of turns. Next, slacken the two upper fixings or remove them altogether depending on what is necessary to move one wheel in relation to another. Sometimes the job is made easier by temporarily removing the fly, making the necessary adjustments then replacing the fly again on completion.

If the plates are pinned, loosen the pins in the bottom two pillars and remove the pins in the top pillars altogether. Do remember to secure all fixings, screws, nuts or pins on completion of adjustments. To assist in getting the right tooth engaged with the correct pinion leaf, and because wheels often turn as the plates are opened, mark with a suitable marker pen the appropriate tooth to be associated with a particular pinion leaf. On completion, remove your identification marks.

8 ALTERNATIVE TYPES OF STRIKING CLOCK

FRENCH STRIKES

French strikes are better quality clocks which may be rack or locking plate. The movements are often fitted to heavy marble or slate cases, and may have the escapement planted in front of the dial when it is known as a 'visible escapement'. Extra careful handling is necessary when assembling these clocks, especially when adjusting one wheel in relation to another to get the striking train to operate correctly, because the pivots are long, thin and hard and can break very easily. Repivoting or making a new arbor and pinion can be difficult.

To assist in setting up the strike train, a separate bridge or cock is fitted to accept the back pivot of the hammer-lifting wheel, which means that the hammer-lifting wheel may be moved independently to get the right tooth engaging with the right pinion leaf of the locking wheel.

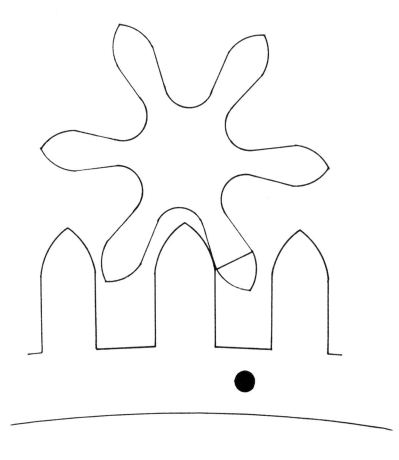

Fig 82 Identification marks to associate a particular tooth space with a particular pinion leaf.

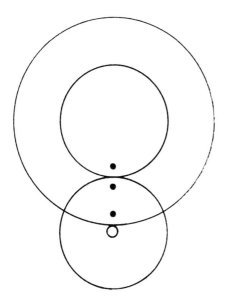

Fig 83 The correct alignment of assembly marks on the cannon pinion, minute wheel and hour wheel of a rack strike.

With the French strike, there is a small countersink to be found between two teeth on the hammer-lifting wheel and a chamfer on one pinion leaf of the mating locking wheel. If the countersink and chamfer are aligned when assembling the movement, the operation between the two wheels should be correct. Similar marks are found on the locking wheel and warning wheel pinion, and correct alignment here should also leave the two correctly associated.

If the strike is a rack type, the cannon pinion, minute wheel and hour wheel will also have countersinks which should be set up so that they are in line. This should ensure that the rack tail falls onto the snail in the right place. The snail of a French strike is smooth, not stepped.

Assemble the movement with marks aligned, then test that the set-up is right. The correct functioning of the train takes priority over any marks. Ideally, when the hammer arbor falls from a lifting pin, the pin in the locking wheel should be close to the locking detent. Once locking occurs, the pin in the warning wheel should be opposite – that is, 180° away from – the warning detent.

If the mechanism is a locking-plate type, the locking plate will be mounted on a square which is an extension of the intermediate wheel back pivot. Expect to find the brass collet on which the locking plate is fixed to be marked for correct association with the square on the intermediate wheel which will also have one flat marked.

On completion of striking the hour, the counting lever should be positioned as seen in Fig 85a. On completion of striking the half-hour, the counting lever should occupy the position shown in Fig 85b.

Fig 84 The ideal position of the pin in the locking wheel in relation to the locking detent when the hammer falls from a lifter.

Fig 85 The position of the counting lever with the train locked after the hour (a), and locked after the half hour (b).

counting lever

locking plate

a

b

Usually a French strike or timepiece has a Brocot-style regulator. This is a block, mounted on the back of the pallet cock, which carries the suspension spring arranged in such a way that its effective length can be adjusted by turning a rod through a small hole at the top of the dial. The winding key should be double-ended, one end for winding the clock, the other provided with a small square to fit the regulating arbor.

The whole movement and dial is fitted to the case and held by two brass strips extending to the back bezel; this has two screws passing through to reach these strips. After getting the dial upright and securing the movement, the clock is put into beat (as explained in Chapter 4, 'Testing for Beat'). The case screws should be sufficiently tight to prevent the movement from turning during winding, yet should allow the bezel to be turned for beat-setting.

Overhauling a French Strike

Start by examining the movement in the usual way to establish why the clock has stopped. Next, identify any holes that need bushing then, after letting the power off both trains, strip the clock and effect any necessary repairs including bushing, pivot polishing and inspecting the ends of both mainsprings. At this stage the movement is polished by brushing the brasswork with a stiff bristle brush (but not too stiff) dipped in brass polish. The clock is then cleaned (personally I prefer paraffin for this, as it does not affect the polish) and dried, and then, handling it in tissue paper, it is French chalked.

This treatment may not be consistent with conservation, but consider the following points: first, the clock has almost certainly already had this treatment a number of times over the past one hundred years so we are not destroying the present surface finish. Second, history is still being made, and what we are doing today with clocks is likely to be of as much interest in three hundred years time as methods of production three hundred ago are to us today. Third, most customers expect their French carriage or four glass clock to be polished, and livelihoods depend on giving customers what they want (within reason). Fourth, if polishing is unacceptable to you, just cleaning the clock in mild ammoniated fluid and then rinsing should produce a very acceptable finish.

After polishing and drying, rub a clean, moderately stiff brush across a block of French chalk, then brush the brasswork leaving it bright, without blemish. Now peg the holes until the pegwood comes out of the holes clean. It is essential that all traces of abrasive material – brass polish and chalk – are removed from all surfaces, including pinion leaves and wheel teeth. Failure to remove all traces of abrasive material will lead to very badly cut pinions in a matter of months.

The hammer arbor, arbor the of the rack hook and locking detent, the clicks, click springs and screws were originally left dark blue from heat treatment. If these parts are in good condition they may be left, but if badly marked they should be improved and then re-blued using a copper bluing pan filled with clean bran filings. Damaged screws should not be turned but may be improved by stoning and burnishing, then heat-treating to restore the dark blue finish.

When replacing the gathering pallet, its tip should point straight up to ensure that it clears the rack when all the teeth are gathered up, and that there is no possibility of it preventing the rack from falling by entering between teeth during the warning.

AMERICAN STRIKING CLOCKS

American striking clocks are always locking-plate type with a riveted loop mainspring. There is no barrel. The counting lever and locking detent do not rely upon gravity to fall, but are spring-assisted. The motion work is always to be found between the plates in the form of 'idling gear' driven by the second wheel. With the exception of the motion work, the pinions are not solid but take the form of pins called trunions, working between shrouds. These pinions are called lantern pinions.

The escapement of American clocks is usually recoil, and may be between the plates or mounted outside the front plate. For this, the escape wheel has a longer arbor and a cock fastened to the front plate to take the front scape pivot. The

Fig 86 The Ansonia American strike.

pallets are mounted on a pin in a bar also fastened to the front plate, and it is moved to adjust the depth of the pallets with respect to the escape wheel. Occasionally American clocks have a Brocot-style escapement and Brocot-style regulator, and some even have stop work fitted. There were a number of makers of American clocks, including the Ansonia Clock Company of New York, Seth Thomas and Gilbert.

American strikes are very similar so I shall explain just one, an Ansonia; however, be prepared to note the differences if you tackle another American strike.

The Striking Train

The great wheel drives the second wheel pinion which has the wheel riveted to the front shroud. In front of the wheel is a second pinion, made of brass, which drives the locking plate.

The second wheel drives the third wheel pinion. The third wheel has a disc riveted to it which has two notches in the rim and two pins pointing towards the front plate. The notches allow the locking detent to fall, but don't lock directly. The two pins act as hammer lifters so at the half-hour, this wheel makes half a revolution.

The next is the fourth wheel, which also acts as a combined locking and warning wheel. The fly is driven by the fourth wheel, and its only function in this clock is to control the speed of the striking train.

To understand how the striking works, it is best to have a similar clock in front of you to study as you read through the following. So, let us assume that the clock has just finished striking an hour – any hour. The counting lever must be in the centre of a deep gap in the locking plate, the hammer will have dropped, and lever 'A' must be in the gap on the disc of the third wheel. The combined locking and warning wheel will be locked by the locking detent. The centre arbor turns, and one of two pins lifts the lifting piece which, in turn, lifts a bent wire secured to the arbor that carries the locking detent. Once released, the combined locking and warning wheel makes almost a half turn, then the pin is arrested by the warning detent.

Fig 87 The relative positions of the counting lever, hammer arbor, the lever that allows locking and the locking detent and warning wheel.

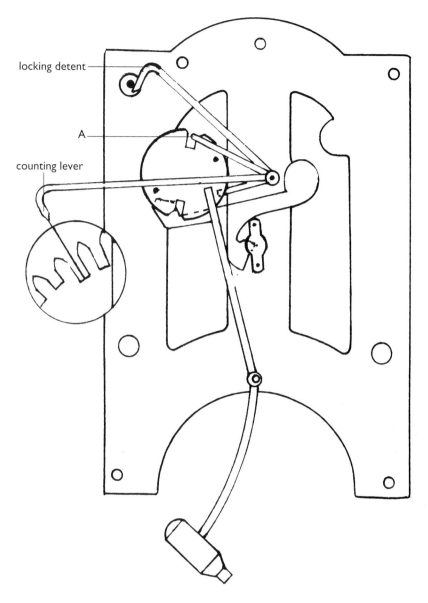

locking detent

A

counting lever

Once the lifting piece drops off the lifter, the third wheel makes half a revolution, allowing one blow to be struck. The locking lever drops into the next deep gap and the train is allowed to re-lock. At the hour, the counting lever drops between teeth in the locking plate, the teeth having shallow gaps that will not allow the train to re-lock; so blows will continue to be struck until the counting lever falls once more into a deep gap.

Before stripping one of these clocks, first make sure that the diameter occupied by the mainspring is just less than the diameter of the great wheel; then tie string around the mainspring to restrain it (this may involve winding it, or letting it down). Once tied, let it down further until all the power is off the great wheel and onto the string – only then is it safe to strip the clock. Remember to take the tension off the hammer arbor and locking detent springs before stripping the clock.

When reassembling the clock, the string must be in place to restrain the mainspring and hold it at about the diameter of the great wheel, otherwise it will not be possible to get the clock back together. Try to assemble it with levers and wheels in their correctly associated places, but be prepared to move one wheel in relation to another to get the striking correct.

Occasionally a spring on a locking detent may break. You can make a new spring as follows:

1. Select 15cm (6in) of spring wire similar to that which was used originally.
2. Mount a piece of steel in the lathe of similar diameter to the arbor the spring is being made for; have about 15mm (⅝in) overhang.
3. Bend the wire near its end to form a right-angle, and insert about 15mm (⅝in) of wire into the split in the collet.
4. Hold the other end of the wire in pliers pulling lightly on the wire while turning the headstock of the lathe by, say, two turns.
5. Now pull harder on the wire while turning the lathe by hand until the required number of turns is achieved. Keep the coils touching by letting the wire slip off the previous coil.
6. Remove your new spring from the lathe and feed it onto the particular arbor it was made for.

On some American clocks, the great wheel is also the hammer-lifting wheel. The fly may double as a warning wheel on some clocks.

TING TANG STRIKING

The ting tang is a type of rack striking which gets its name from two hammers operating on two different pitch gongs at a quarter past the hour,

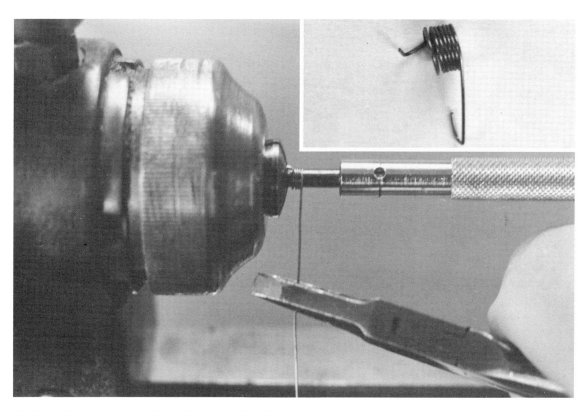

Fig 88 Making a new spring for an American strike. The finished spring (inset).

at half past and a quarter to the hour. At the first, second and third quarters respectively there will be one, two and three sequences of two hammers on two gongs. At the hour, the hammer of the higher note is held in an elevated position while the hour only is struck on the lower note, without the ting tang.

When assembling and setting up the strike, follow the usual procedure setting up the strike according to the arbor of the lower note in relation to the hammer-lifting wheel, ignoring the arbor of the higher note which should automatically fall correctly.

For the operation of a ting tang, some familiar parts will need to be modified. For example, whilst the majority of the teeth of the rack will have tips on an arc of a circle struck from the pivoting point of the rack, the first three teeth will be struck from a different centre, effectively giving tips at different distances from the pivoting point of the rack.

The cannon pinion has four lifters, each at a different radius from the centre, to control how many teeth of the rack fall. When the first pin lifts at a quarter-past, only one tooth of the rack falls

and we get one sequence of ting tang. At half past there is a higher lift, allowing two teeth to fall and two sequences of ting tang. The third quarter is triggered by an even higher lift allowing three teeth of the rack to fall and the ting tang operates for three sequences.

At the hour, a pin in the minute wheel causes a lever to lift the arbor of the higher note's hammer so that it cannot be lifted by the hammer-lifting wheel; consequently only the hour will be struck. The hour is triggered by the pin furthest from the centre of the cannon pinion.

The strike snail offers special problems with this type of strike in that, for half-past one and a quarter to two, more teeth of the rack must fall than a normal snail would allow. To overcome the problem, the one o'clock step of the snail is specially shaped allowing only one tooth of the rack to fall at one o'clock, two teeth to fall at half past one, and three teeth at a quarter to two. At a quarter to three the rack tail may fall onto the three o'clock step thus allowing three rack teeth to fall. Careful attention is necessary when positioning the hour wheel of a ting tang.

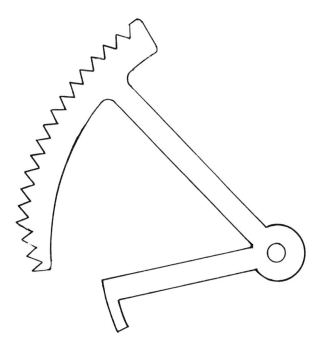

Fig 89 Rack of a ting tang strike.

REPEATING CLOCKS

Probably the most common type of repeater encountered will be the French carriage hour repeater. You may come across a quarter repeater, a five-minute repeater or even a minute repeater. Here, only the hour repeater will be described. They are easily recognized by a repeat button positioned at the top of the case: when the button is fully depressed and released, the last hour that struck is repeated.

On a normal rack strike, the hour snail is fixed to the hour wheel, but on a repeater, the hour wheel is fixed to a star wheel allowing it to remain stationary for most of the hour, advancing just a few minutes before the hour by a pin in the front of the cannon pinion. The star wheel has a jumper engaging between two teeth to hold the snail stationary and to permit a complete number of teeth of the rack to fall, minimizing the chance of the gathering pallet butting a tooth tip in the rack. The step in the snail is cut vertically, or is undercut for the same reason. It is only during approximately the last three minutes of an hour that the next hour will be struck on operation of the repeat lever, rather than the last hour.

So that the repeat lever can be operated at any time, there is no warning on a repeating train. Instead, when the repeat button is depressed, a pin in the repeat lever engages the fly. The fly is released, and thus the train, when the repeat lever is released, and this ensures that the correct number of blows is struck.

With the exception of the warning wheel, the strike train, hammer arbor and locking lever work in a similar manner to the ordinary French strike.

Under normal operation the 'lifting piece' is pulled back against its spring by one of two pins in the cannon pinion. Unlocking lever 'A' is also pulled back, and the end with the step falls so that

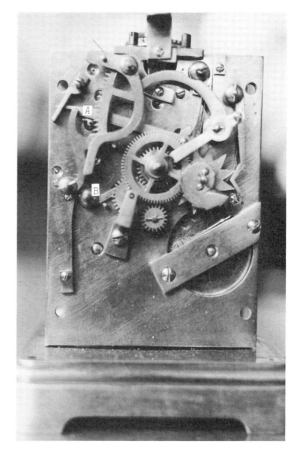

Fig 90 The star wheel and snail on a repeater clock.

upon release it is fired forwards and the vertical face pushes the rack hook back so that the rack falls onto the snail and the train is unlocked. The teeth in the rack are gathered up one at a time until the rack hook drops significantly deep for the locking detent to lock the locking wheel.

At the half hour, a pin in the minute wheel engages a blocking lever 'B'. This prevents the rack from falling by another pin protruding from the back of the rack coming to rest on the blocking lever.

9 CHIMING CLOCKS

A chiming clock has three trains: a strike train, normally on the left when looking at the clock dial; a time train in the middle; and a chime train on the right. The chime train plays a tune at each quarter and on the hour, the most usual being the familiar Westminster chime. There is a sequence of four hammers at a quarter past the hour, eight hammers at half past the hour, twelve at a quarter to, then sixteen hammers at the hour followed by the hour strike. The correct time should be indicated by the first hammer blow of any hour after the tune.

There are, of course, only four hammers used on the Westminster chime, operating on four rods or gongs of different length; the notes used are G sharp, F sharp, E and B. Sometimes there is a fifth gong and hammer which is used to indicate the hour only, to produce a pleasing tone. Other tunes are used, those met the most frequently being Whittington and St Michael which operate on a full octave of eight different notes. On some clocks all three tunes are available on the same movement, a particular preference being selected by the slide of a lever. The same lever, or a separate one, may give the facility of silencing the chime when it is not required. Other clocks have a feature where the chime automatically 'cuts out' at night and cuts back in again in the morning.

Most chiming clocks have removable barrels which come out without the need for stripping the clock. There are several different arrangements to allow this, all self-explanatory.

The following discussion is an overview of how chimes work in general, and then the Smiths K6 in particular. This is followed by detailed instruction for dismantling and reassembling. What is learned should be transferable to other chiming clocks. The part names used are those which are more generally recognized; they are not necessarily Smiths' own part names.

CHIMES IN GENERAL

The link between the time train and the chime train is through one of four lifters on the cannon pinion, one having the ability to lift the lifting piece higher than the other three. The strike train is initiated into operation by the chime train.

In a self-correcting chime, at the end of chiming the hour and the first and second quarters, the chime train locks on just one locking detent known as the 'single locking detent'. At the end of the third quarter, two detents lock the chime train, the second one being known as the 'double locking detent'. Only the high lifter on the cannon pinion will unlock the chime warning wheel, by lifting both detents; it thus puts the chime sequence into step with what the hands indicate, should they have become out of step – for example when adjusting the hands.

Whilst half a wheel warning remains the same on the strike side of a chime clock, a quarter of a wheel warning is necessary on the chime train.

When the last hammer falls at any quarter, the chime train should run on by one quarter to one half of a turn of the warning wheel. This is just a safety margin to allow for wear and small inaccuracies in manufacture.

HOW THE SMITHS K6 OPERATES

Let us assume that the strike and chime train are locked, and that the first low lifter on the cannon

Fig 91 The first low lifter is about to lift the lifting piece. (The hour wheel is removed for clarity.)

pinion is about to lift the lifting piece and chime warning detent. The cannon pinion continues to lift until a pin in the chime warning detent contacts the strike warning lever, the arbor of which carries the chime double and single locking detents. Eventually the chime train is unlocked and the warning wheel advances one quarter of a turn to the chime warning detent.

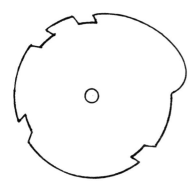

Fig 92 The chime locking plate.

The lifting piece continues to lift briefly, but without releasing the strike train. Eventually the lifting piece drops off the lifting cam.

The chime locking wheel makes one revolution, and then, as the strike warning lever is allowed to drop because of the gap in the chime locking plate, the chime train, having given a sequence of four hammers, re-locks on the chime single locking detent.

Something similar happens at half past the hour, except that the chime locking plate will not allow the chime train to re-lock until the locking wheel has made two revolutions and eight hammers have fallen (two lots of four). At a quarter to the hour, three revolutions of the locking wheel occur; then, due to the chime corrector which is attached to the back of the chime locking plate, locking occurs not only on the single locking detent but on a second detent called the double locking detent.

At the hour, both detents are lifted by the high lifter, but partway through the sequence a high spot on the chime locking plate lifts the strike warning lever sufficiently high to lift the rack hook on the striking side and allow the half wheel warning on the strike to occur. The striking train is not freed until the strike warning lever falls by virtue of the pin in the lever falling into the step in the chime locking plate. The chime corrector will have turned, so locking of the chime train is once more by the single locking detent only.

For a brief moment at the end of the chime sequence both chime and strike train will be operating at the same time, but this will not be noticed by the owner of the clock. It is only noticed when testing the strike and time train while controlling the speed of the chime train. The operation of the striking train is sufficiently similar to what has already been described that no further explanation should be necessary, except to say that often striking occurs on three gongs giving a chord and pleasant tone.

DISMANTLING

Having identified the cause of stoppage and the repairs that are necessary; proceed as follows:

1. Let the power off all three trains. Test by feeling the freedom of the barrels before stripping.
2. Remove the suspension spring, rod, pallet cock and pallets.
3. On the front of the clock remove the rack, chime/silent lever, minute wheel, hour wheel, rack hook, lifting piece and chime warning detent.
4. Unscrew the small screw holding the chime locking plate and remove. It may be necessary to turn the train by hand to gain access to the screw.
5. Unscrew the hammer-lifting rod assembly and lift the rod from the hammer arbor.
6. Remove the complete hammer assembly by removing two screws from the front plate and two from the back.
7. Remove the one remaining screw from the back of the hammer assembly and, having taken off the small plate, remove the chime barrel ('chime barrel' being the name given

to the assembly of four or sometimes eight lifters). It is not essential to dismantle the hammers but if it *is* felt necessary – to remove congealed oil, for example – do write down the sequence of hammers, bushes or washers for reassembly after cleaning.

8. Remove all three ratchet wheel bridges, ratchet wheels, bushes and all three barrels. Remember to mark which is which and mark for dismantling the barrels.
9. From the back of the movement, remove the large ratio wheel and the two smaller ratio wheels. Note that the lower one has a smaller diameter pipe.
10. Remove the back plate carefully after unscrewing three nuts and one screw.
11. To aid reassembly, it is usual to mark each wheel discreetly according to the train to which it belongs. Striking train wheels are marked with an 'S', time train wheels are marked with a 'T' and chime train wheels are marked with a 'C'. Some wheels are so similar that a mix-up could easily occur. Mark all wheels on the same side, which will show which way up each wheel goes back into the plate. Some repairers number the wheels to help reassembly.
12. Remove all wheels, both flies and the hammer arbor.
13. Remove the gathering pallet and gathering pallet wheel with suitable support and an appropriate pin punch.
14. Remove the single and double locking detent and strike warning lever.
15. If the centre wheel front pivot and hole need no work, the cannon pinion and centre wheel need not be removed.

The clock is now completely dismantled, and the work that is necessary should be carried out. Clean the clock and peg the holes for reassembly.

REASSEMBLY

1. Reassemble, and lubricate all three barrels with heavy clock oil.

2. Lubricate with heavy clock oil the friction clutch on the centre wheel and the wrap-around pin holding the spring and wheel, and try to get some oil between the inside of the pinion and the centre wheel arbor. Test the adjustment of the clutch arrangement.

3. Replace all wheels into the bottom plate, also both flies and the hammer arbor, checking that all pivots are in good condition and straight. Check that the wheel teeth are undamaged. Do not replace the barrels yet.

4. Replace the back plate manipulating each pivot into its hole, then secure the back plate with its three nuts and one screw. As usual, hold the clock with your thumb under the front plate and fingers resting lightly on the back plate, and manipulate the wheels with your tweezers held in the other hand.

5. Once the plate is secured, hold the movement horizontally and lift each wheel in turn and see if they drop again under their own weight when released. Do the same with the other plate at the bottom: all wheels and arbors should lift and drop under their own weight, and if one doesn't, investigate. As previously stated, the usual faults are bent pivots and distorted plates.

From here on, the golden rule is to replace the minimum necessary at a time, and to ensure that all functions are correct before moving on to the next step. Don't be tempted to replace everything, and then try to sort it out.

6. We are now going to test that with the chime locking wheel locked, there will be one quar-

Fig 93 The lifting piece, chime warning detent and spring.

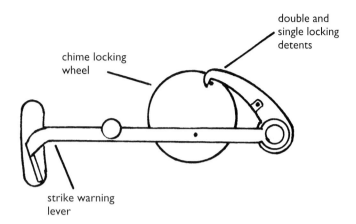

Fig 94 The relative positions of the strike warning lever, the double and single locking detents and the locking wheel.

chime locking wheel

double and single locking detents

strike warning lever

ter of a wheel warning. For this test, replace the chime warning detent first and then the lifting piece, and secure. Lubricate the post they fit onto first with medium oil. Check for endshake. Check that the chime warning detent falls fully, and that there is a minimum gap between the pin in the lifting piece that lifts the chime warning detent, and the detent itself. This will ensure that the hands can be turned backwards safely and that the lifting piece returns to its correct position. Any necessary adjustments are made on the return spring.

Now, replace the strike warning lever, and the double and single locking detents. Turn the chime train by hand until both detents lock fully onto the chime locking wheel. With the strike warning lever held fully down, 'tack' down the screw in the single locking detent leaving safe endshakes (I would consider 'safe' to be between $^{10}/_{100}$mm and $^{20}/_{100}$mm).

7. With the chime locking wheel locked, check where the warning wheel pin is in relation to the warning detent. Looking from the front of the movement, the warning wheel turns clockwise so the pin in the warning wheel needs to be 90° from the warning detent. For this test, lift the double locking detent by hand or you might not get a true picture because the two locking faces are not accurately aligned.

If, with the single locking detent locked, the chime warning wheel is not in the right place, slacken the pillar nuts enough to move the warning wheel in relation to the locking wheel to bring it right. Be careful not to lean the warning wheel too far, or a pivot will get bent or broken; open a minimal gap between the locking wheel teeth and the leaves of the warning wheel pinion. Once the warning is correct, secure the plates again.

When working on a chime which has the chime corrector between the plates on the chime third wheel, make sure that when the corrector comes into play, which it will do with every ten revolutions of the locking wheel, the chime corrector lever rests safely in the low part of the chime corrector. If necessary, both the chime locking wheel and the chime warning wheel will have to be adjusted to co-ordinate all three.

8. Replace the chime barrel bush, ratchet wheel and ratchet wheel bridge, lubricating all friction points with heavy oil. Wind the chime mainspring by about one third (six half turns of the winding key).

9. With the chime train locked, replace the chime locking plate giving access to the securing screw and so that the pin in the strike warning lever is in the position shown in Fig 96. That part of the double locking detent which rides on the chime corrector cam will have to be lifted to replace the chime locking plate.

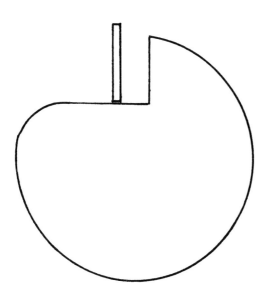

Fig 95 The chime corrector lever resting safely in the chime corrector, with a gap between the vertical face and the lever.

When working on a chime with the chime corrector on the chime third wheel, cause the chime corrector to operate, then replace the chime locking plate so that the pin in the strike warning lever is in the third gap of the locking plate – that is, as if the hour is about to chime.

Now we are going to check the chime corrector to ensure that only the high lifter will release the chime train when the chime corrector comes into play, also that the unlocking is safe, and that we have safe locking.

10. Cause the chime corrector to operate by turning the cannon pinion. For this, put your index finger on the escape wheel to stop it

from turning, and keep it there for all the chime checks. Now, with the chime corrector operating, check the gap between the tip of the double locking detent and the pin in the locking wheel when the lifting piece is at the very top of the high lifting cam; a safe clearance should be about $\frac{5}{100}$mm. If the gap is not correct, adjust by altering the angle between the single locking detent and the strike warning lever. The angle is easily altered because

Fig 96 Correct position of the pin in the strike warning lever relative to the chime locking plate.

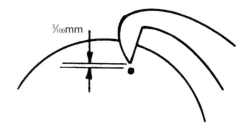

$\frac{5}{100}$mm

Fig 97 When the chime corrector is operating and the high lifter is at the peak of its lift, there should be a safe clearance between the locking pin and the double locking detent.

the securing screw of the single locking detent was only tacked down.

When the clearance is right, tighten the screw in the single locking detent a little more, but not fully yet, because all tests are not completed.

Still holding the chime fly, turn the cannon pinion clockwise until the lifting piece drops off the high lifter. Now, briefly releasing the chime fly, turn the chime train backwards to re-lock both detents. (Alternatively let the chime train operate until the chime corrector comes into play again, then turn the cannon pinion backwards to the first of the low lifters.)

11. Still with your finger on the escape wheel, turn the cannon pinion forwards so that the first low lifter tries to unlock the chime train. It should remain safely locked on the double locking detent.

Try the next two low lifters. Still the double locking detent should hold up the chime train. Now try the high lifter again, and the chime locking wheel should unlock giving safe clearance between the pin and the double locking detent. Allow the train to run: it should re-lock on the single locking detent only after four turns of the chime locking wheel.

12. Turn the cannon pinion so that once again the first low lifter unlocks the chime train. Control the fly and stop the train so that the locking pin is directly under the single locking detent. Observe the gap between them, then without releasing the train, try the other two low lifters. If all is well, tighten the screw in the single locking detent fully and check that nothing has moved.

13. The final check on the chime is that locking is safe on all four quarters; that when the chime corrector operates, only the high lifter will unlock the chime train.

Now you can release the escape wheel and give a welcome rest to your index finger.

Reassembly of the Strike Work
1. Lubricate the rack hook post with medium

oil and replace the rack hook. Check for endshake.

2. Lubricate and replace the strike barrel, bush, ratchet wheel and bridge. Wind the strike train.

3. Bring the lifting pin in the hammer arbor between two lifters, then allow the train to operate while putting a light turning force on the hammer arbor to simulate the weight of the hammer. Controlling the fly and holding the rack hook up, stop the strike train immediately the hammer arbor falls from a lifter. Now observe where the strike warning wheel pin is. Allow the warning wheel a quarter of a turn, plus whatever turn is necessary (it may be none) for the pin in the warning wheel to come to rest against the strike locking detent which is part of the rack hook.

4. Now replace the gathering pallet so that the projection which rides on the periphery of the gathering pallet rests to the right of centre of the gap. Secure the gathering pallet by supporting around the back pivot of the gathering pallet wheel with a flat hollow punch, driving the gathering pallet on firmly with a cannon pinion punch.

5. Replace, in order, the minute wheel, hour wheel and rack. Turn the cannon pinion while once again holding the escape wheel stationary until the high lifter operates. Immediately the lifting piece drops off the high lifter, lift the rack hook and adjust the hour wheel so that the rack tail falls on the centre of a step on the hour snail. Release all trains and allow the teeth of the rack to be gathered up.

6. Replace the washer on the minute wheel post; secure and check the endshake of the minute wheel.

7. Replace the time barrel, bush, ratchet wheel and ratchet wheel bridge, and secure. Do not wind the time mainspring.

8. Reassemble the chime hammer assembly. Replace the chime barrel (lifting cams) and replace the small back plate, securing it with its one screw.

9. On the back plate, replace the second and

third ratio wheels, remembering that the one with the thinner pipe goes on the bottom post.

10. Replace the chime hammer assembly, securing it with the four remaining screws.

11. Place the strike hammer lifting rod assembly over the hammer arbor, adjusting the angle so that there are about three turns of the warning wheel between the hammer falling off one lifter, to its starting to lift by the next lifter.

12. Cause the chime locking plate to indicate that a quarter past the hour has chimed. Simply lift the strike warning lever to do this until the pin on the strike warning lever drops into the gap in the locking plate.

13. Replace the large chime ratio wheel, lightly tacking down one securing screw (it may have two).

14. By looking at the chime gongs, discover the sequence for the hammers to chime a quarter past. The shortest gong should be struck first, followed by the second shortest, then the third shortest and finally the longest gong.

15. To get the hammers agreeing with the locking plate, turn the large ratio wheel keeping the chime train locked until the correct sequence is found. Once found, tighten the screw in the ratio wheel just a little more so that, if necessary, fine adjustments can be made by again turning the ratio wheel on its post (the chime third wheel back pivot).

16. Now we are going to make any fine adjustments. Holding the escape wheel stationary as before, release the chime train allowing

seven hammers to operate (for a Westminster chime), then control the falling of the eighth hammer by means of the fly. When the eighth hammer falls, stop the chime train immediately and observe where the warning wheel pin is. Now allow the warning wheel to advance, under control, until the single locking detent locks the locking wheel.

Ideally the warning wheel will advance by one quarter to one half of a turn, from the last hammer falling from a lifting cam to locking of the chime train. If it is more, or less, make fine adjustments by turning the large ratio wheel on the chime third wheel. Usually adjustments amount to the equivalent of a small part of a tooth.

After a fine adjustment, it will be necessary to move on to the next quarter. Keep this up until, on all four quarters, the chime runs on by between one quarter and one half of a turn of the warning wheel.

Should you wish to know the sequence of hammers for any other quarter, they are shown in Fig 98. For convenience I have numbered the hammers 1 to 4, with No 1 hammer striking the shortest rod and No 4 the longest.

If the chime has more than one tune, it will be necessary to check the safe action on all tunes and all quarters. With an eight hammer chime, it is often prudent to go for just a quarter of a turn of the warning wheel when the last hammer falls. Significantly less than a quarter of a turn on the warning wheel when the last hammer falls, and the chime may be made unsafe.

First quarter	Second quarter	Third quarter	Fourth quarter
1 2 3 4	3 1 2 4	1 3 2 4	3 1 2 4
	3 2 1 3	4 2 1 3	3 2 1 3
		1 2 3 4	1 3 2 4
			4 2 1 3

Fig 98 The sequence of hammers for each quarter. Number one is the hammer for the shortest gong, number four is the hammer for the longest gong.

The object of all this is, of course, to make sure that all the hammers do fall at each quarter, but also that no hammers are on the rise during the warning.

The first quarter on both Whittington and St Michael is 1 2 3 4 5 6 7 8 assuming, as before, that the hammers are numbered 1 to 8, starting with the hammer that strikes the shortest gong.

17. Tighten fully the screw in the ratio wheel when all fine adjustments are made. If there are two screws, tighten only one fully. The other is only lightly secured to stop the screw falling out. Two screws are mechanically unsound though they will often be found.
18. Replace the pallets, pallet cock and screws. Wind the time train and adjust the pallets.
19. Replace the suspension spring and suspension rod.
20. Replace the chime/silent lever.
21. Lubricate now, if not done as work progressed, the larger heavily loaded pivots with heavy oil, and both flies, warning wheels, scape wheel and pallets with medium oil. Lubricate warning pins, locking pin and gathering pallet pin with medium oil. Lubricate all points of friction with heavy or medium oil as appropriate, remembering to oil the tails of all hammers where they rest against their stops. Failure to do this will cause a 'jerky' chime that may stop the chime train operating.

    Finally, case the movement, wind all the trains fully, and test.

ALTERNATIVE CHIME-OPERATING SYSTEMS

On another Smiths, the chime fourth wheel acts as both locking wheel and warning wheel, even though there is another wheel between it and the fly. At the end of chiming each hour, the chime fourth wheel is locked by the locking detent falling into the gap in the cam.

To unlock the train for the next hour, the lifting piece lifts the strike warning lever which carries the locking detent. By the time the wheel is unlocked, the warning detent – which is on the same arbor as the lifting piece – has risen into the path of the warning pin. Next, the lifting piece drops off the lifting cam on the cannon pinion, therefore the warning detent drops, freeing the pin and wheel.

The locking wheel will make one revolution for the first quarter, then it re-locks because another projection on the strike warning lever falls into a gap in the chime locking plate; similarly for the half past, except that the chime locking wheel will make two revolutions.

At a quarter to the hour, the chime locking wheel will make three revolutions, but just before locking occurs, a stepped lever in the back of the chime locking plate contacts a 'D'-shaped pin on the lifting pin, tensioning a spring. If any of the three low lifters now lifts the lifting piece, this will lift high enough for the chime corrector to advance so that the 'D'-shaped pin will rest on the step in the chime corrector; but as the lever doesn't advance sufficiently, the warning detent can't drop to clear the warning pin.

At the hour, the high lifter lifts the lifting piece sufficiently high for the chime corrector to spring forwards so that when the lifting piece clears the lifting cam on the cannon pinion, the warning detent falls free of the warning wheel.

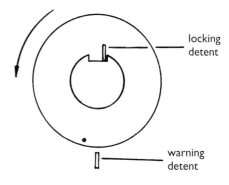

Fig 99 The system for locking and warning on an alternative Smiths chime.

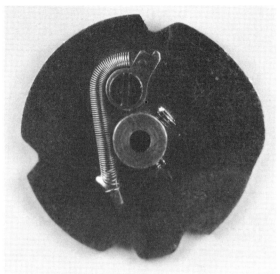

Fig 100 An alternative chime corrector before tensioning (left) and after tensioning (right). The 'D'-shaped pin is not shown.

10 WEIGHTS, SPRINGS, STOP WORK AND THE FUSÉE

The significant advantage of a weight as a driving force for a clock is its constant output of power throughout the clock's running period; this advantage gives a more constant impulse, resulting in a more constant amplitude to a pendulum or balance and thus better time keeping potential. The two great disadvantages with a weight are first, it needs space in which to fall, and second, weight-driven clocks are not portable. The discovery and development of the laws of springs in 1665 by Dr Robert Hooke led to the mainspring being used as a driving force, and the spring to do for a balance, what gravity does for a pendulum.

The power curve of a mainspring initially describes a steep drop-off in power, followed by a more linear and gradual drop for most of the running period of the clock, ending in another sharp drop-off. The effect of this is a similar variation in impulse, and therefore amplitude, so that, due to escapement functions, a clock working with a balance can be seen to lose with a reduction in amplitude. The maximum effect of the change in power as a mainspring runs down can be reduced if the top and bottom parts of the spring are not used. This was done by fitting what is known as 'stop work'.

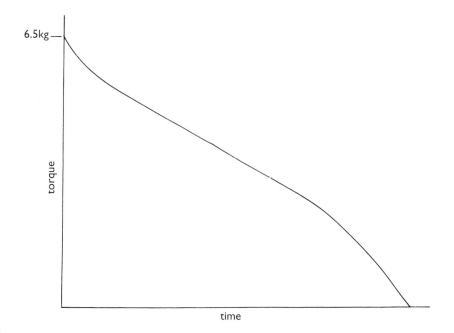

Fig 101 The power curve of a clock mainspring.

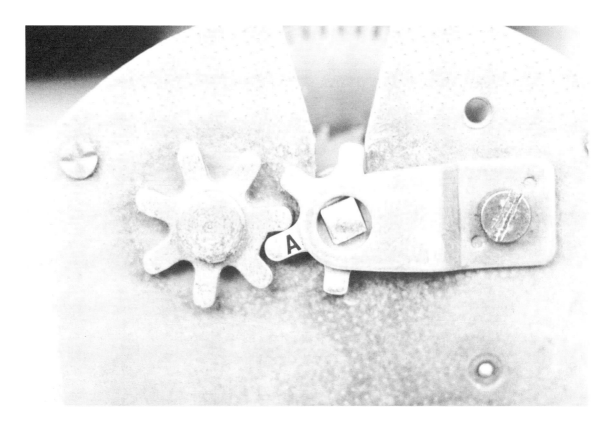

Fig 102 Stop work (upper and lower).

STOP WORK

Stop work can be thought of as a pair of gears, one fitted to the barrel arbor, the other to the barrel directly, or to the plate. The pair of gears takes the form of star wheels, or what is often called a Maltese cross. The star wheel fitted to the square on the barrel arbor has six teeth, one of which is longer than the other five. The other star wheel has seven teeth, all of the same length, but the root between two of the teeth is shallow so that the long tooth can never pass through the shallow space.

There are seven turns of the winding key to wind the clock fully from a completely run-down state, so the star wheels are arranged to use the middle part of the mainspring avoiding the two extremes of a fully wound and a completely run-down mainspring. In this instance, the clock is an eight-day so should be wound every seven days, although it could in fact run for about eight and a half days. In the illustration, the stop work is shown with the mainspring in the fully run-down state. Tooth 'A' is coincident with the shallow space in the second star wheel and cannot pass. When winding, the long tooth passes through the deeper gaps until it is once more coincident with the shallow space.

Before stripping a clock with star-wheel or Maltese-cross stop work, consideration must be given to removing the 'pre-load', that is, the power still on the mainspring when the stop work locks to prevent the last bit of the mainspring being used.

Taking Off the Pre-load

In the first example given, the bridge holding the star wheel with the long finger is removed, followed by the removal of the star wheel itself. The clock is now run down under control by removing the platform if fitted, or by letting it down on the click.

In the second example, where the stop work is fitted to the barrel, let the power off in the usual way, then remove the barrel from the clock but don't dismantle the barrel yet. Instead, hold the winding square in the soft jaws of a vice, turn the barrel slightly to take tension off the stop work, and then either unscrew the Maltese cross from the barrel cover, or slide the piece with the finger up the arbor then let the pre-load off by allowing the barrel to turn in the normal direction of drive. There is likely to be about half to two turns of the barrel.

Setting up Stop Work

The method I would choose to set up stop work and the amount of pre-load I would put on is likely to differ from that used originally. I tend to use a little more of the strong part of the mainspring and therefore a little less of the weak part: this takes account of a mainspring that has been in the clock for many decades and has become tired, and of the general wear and tear in the clock. This will not apply to all clocks, and I would only use the stronger part of the spring where I felt it necessary.

To set up stop work when it is fitted to the barrel arbor and clock plate, with the platform removed but holding up the train, wind the clock fully with the stop work disengaged. Let the clock run for two turns of the barrel arbor, then replace the stop work in such a way as to prevent further winding. Finally, allow the clock to run down until the stop work stops the clock again, but then check that there is still sufficient power to drive the clock, even when it is almost run down. If there is not, adjust the stop work again to the point where there is only half a turn of the barrel arbor before the mainspring becomes fully wound.

When the stop work is fitted directly to the barrel, first reassemble the barrel then hold the square of the arbor in the soft jaws of the vice while turning the barrel by one turn. Secure the stop work to prevent the mainspring running down any further.

Finally, if the stop work is fitted to a clock with a balance rather than a pendulum, check the amplitude of the balance, and adjust the pre-load if necessary to what would be acceptable. A good amplitude is about 270°, but it should not drop below 220° until the stop work operates.

Regrettably there are too many examples of clocks where the stop work has been removed altogether. Whether this is through malfunction, or ignorance of its purpose, or ignorance of how to set it up, I don't know, but it is sad to see.

Stop work can be found on modern clocks, but it is more usually found on French carriage clocks and a number of other quality clocks. There are examples on American strikes as well.

THE FUSÉE

Stop work did offer some advantage to better timekeeping, but the use of the fusée was significantly superior in achieving something very close to the constant force derived from the weight. The principle of operation is that the great wheel has no teeth, and that power is transmitted

Fig 103 A fusée wheel.

instead by gut, chain or wire pulling on a small diameter when the power of the mainspring is greatest, and pulling on a large diameter when the mainspring power is least. To achieve this, the second wheel has a continuous groove running around a fusée barrel, gradually increasing in diameter. The curve of the fusée can be calculated, with the use of calculus and resembles the curve of the thumb with the first joint locked.

The mainspring is given a pre-load, after which the square on the barrel arbor is not used again. Instead, the clock is wound by the square on the fusée wheel; in effect it is the barrel that turns to wind the clock rather than the barrel arbor. When the fusée wheel is full, a stop iron is actuated that prevents the line coming off the end of the fusée wheel. As the clock works, the line pulls on an increasing diameter to compensate for the weakening of the mainspring's pull and so a very near constant impulse is given to the pallet through the centre, third (or fourth) and the escape wheels.

Some fusée clocks have a groove running around the fusée wheel with square corners at the bottom of the groove and straight vertical walls. These are chain-driven. Those with a rounded groove are gut- or line-driven.

DISMANTLING A FUSÉE CLOCK

There are three points worthy of note before work is started on a fusée clock: first, mainsprings on fusée clocks are notoriously strong; second, only exceptionally are mainsprings let down on the barrel arbor; and third, safety glasses are recommended when working on fusée clocks especially those with fusée chains. I know of a clock repairer who had a spectacle lens smashed by a fusée chain parting, and another who had a finger broken by a fusée chain. Wear safety glasses, and keep hands and face away from a direct line with a fusée chain.

1. After removing the clock movement from the case, inspect particularly the third wheel back pivot. Quite commonly this is badly cut, and

evidence of a cut pivot – recognized by red dust around the hole – will mean that the train cannot be run down by removing the pallet.

There are two preferred ways of letting the power off a fusée train: either, remove the pallets and allow the clock to run down; or, let the clock down by way of the click in the fusée wheel. If the maker provides for this, a small hole will be found between two teeth of the fusée wheel which gives access to the tail of the click. If there is a hole, insert a piece of steel wire, wind the fusée wheel just a little, disengage the click and then let the mainspring down by half a turn of the winding key; release the click, then wind by one tooth to ensure that it is engaged in the ratchet wheel. Remove the key and reposition to repeat the operation until the line is completely off the fusée wheel and onto the barrel. (The letting-down process is fully explained in 'Dismantling', Chapter 3.).

If there is no hole, and as long as there are no badly cut pivots, remove the pallets and allow the clock to run down; it may be desirable to oil the pivot holes to assist in this. The clock should run down until the fusée line is all on the barrel and none on the fusée.

If neither of these two methods is possible, there is no alternative but to let the power off by means of the square on the barrel arbor – though let me remind you of the strength of the mainspring. Thus, secure a hand vice, or a letting-down key for preference, on the barrel arbor winding square; make sure that the movement is held securely and cannot spin. Then, holding the vice or key ready to take the strain of the mainspring, carefully slacken the screw holding the great wheel ratchet by half a turn, then disengage the ratchet and let the mainspring down under control. If you wish to stop part-way for any reason, simply hold the barrel arbor stationary and re-engage the ratchet. Repeat this on the strike and chime trains if they are present.

2. If it were possible to let down the mainspring by the fusée click or by removing the pallets

and allowing the train to run down, there is still a pre-load on the mainspring that has to be removed. Follow the instructions given in the preceding paragraph to do this. It should be noted that the fusée click has no click spring and the click screw has no shoulder. This is because once the pre-load is set, the click is locked.

3. When all the power including the pre-load is off, which will be half to two turns of a barrel arbor, strip the clock completely, repair and clean. To strip the fusée wheel, unscrew the screw holding the brass key piece, slide it back and lift free.

Fusée Chain

Clean the chain in clock-cleaning fluid. If it is very bad, rub over it with the same abrasive block that was recommended for cleaning the graver, then clean it thoroughly to remove any abrasive material. To lubricate it, put a few drops of heavy clock oil into a spirit jar that is part-filled with degreasing agent. Put the chain into the mixture for a few minutes, then remove it and allow the agent to evaporate, leaving a thin film of oil over the chain.

Gut Line

If the original line is in good condition it may be used again, though avoid getting kinks in it; just pass it through a lightly oiled cloth to keep it in good condition.

Bronze Wire

Simply wipe over the wire with a clean cloth, inspecting for any split strands. If there are any, renew the wire.

FITTING A NEW LINE

Ideally a new line should fill the fusée wheel and leave about three-quarters of a turn on the barrel when the mainspring is fully wound. This will allow the fusée to work correctly, and the knots to be re-made in a subsequent repair without the need to fit another new line.

*Fig 104 Figure-of-eight
knot in the end of a line.*

*Fig 105 Method of
securing a gut line or wire
to the barrel.*

1. Pass one end of the new line through the hole in the fusée wheel and knot the end using a simple knot. If there is a risk of a gut line pulling through the hole, tie it in a figure-of-eight knot.
2. Reassemble the fusée wheel, lubricating the brass key piece and fastening with its screw. Test that there is no end play – although the wheel must be free.
3. Wind the line around the fusée barrel until it is full; allow enough line to reach the barrel including three-quarters of a turn plus enough to secure the end to the barrel.
4. Secure the end to the barrel. You should now have the barrel and fusée wheel joined by a length of gut or bronze line.

REASSEMBLING A FUSÉE TIMEPIECE

1. Having joined the fusée wheel and barrel with line, replace the barrel, fusée wheel, centre wheel, third wheel (fourth wheel in the train) and the escape wheel into the pillar plate. Make sure that the line is neither kinked nor around a pillar. If the clock has a fusée chain, just assemble it without the chain.
2. Replace the ratchet wheel, click and click screw, but don't tighten the click fully yet.
3. Wind all the line onto the barrel, attempting to position it much as it would lie when coming off the fusée cone. If instead of a line you have a chain, hook one end of it around the pin in the fusée; the other end is hooked into the barrel. (The two ends are different.) Then wind it all onto the barrel as described for a line.
4. Keep pressure on the mainspring until the fusée line is pulling straight from the hole in the fusée wheel, and the escape wheel comes to rest.
5. Ease the pressure on the mainspring but only until there is little or no pressure on the great wheel; do not let the line unravel from the barrel.

Fig 106 Two different ends of a fusée chain. The upper hook attaches to the barrel, the lower to the fusée wheel.

6. Put one turn pre-load on the barrel, then engage the click fully and tighten the screw. Unless the pre-load is to be adjusted, the arbor is not turned any more until the clock is dismantled for overhaul again.

7. Prevent the escape wheel from turning by inserting a piece of pegwood into the wheel crossings; then wind the mainspring by the fusée wheel arbor helping to guide the line onto the fusée groove. This is particularly necessary when a fusée chain is involved, because if the line tried to jump a groove, permanent damage would almost certainly occur to the fusée cone, and a new one would have to be cut.

8. As the line nears the end of the fusée cone, the line will contact the fusée stop iron so that on the last turn, the stop iron locks the fusée wheel.

9. Check that on the penultimate turn of the fusée wheel arbor, the stop iron is safely clear of the hook. Also check that on the last turn they engage safely.

10. Finally, let the train run and observe that there is the same energy in the train when fully wound as when it is almost completely run down.

NOTE: If it is stronger when nearly run down, there is too much pre-load, in which case let it off by one or two teeth of the barrel ratchet wheel and try it again. If it is too weak at the bottom end, put on more pre-load.

It is possible to use a torque bar to check that the power is the same throughout the mainspring, but it is not essential. With the present emphasis on conservation it is not likely that a new fusée wheel would be cut if the power were found to be the same at each end of the mainspring but to vary in the middle.

Using a Torque Bar

Wind the clock fully, but leave the pallet out of the clock. Put the torque bar on the winding square so that it is horizontal and will lift when the escape wheel is released (actually any angle will do). Now observe the action of the bar as the clock runs down – don't let it pass beyond upright. Test that the action of the torque bar is similar at various stages of running down, particularly towards the bottom end of the mainspring, and adjust the pre-load as necessary.

REPAIRING A BROKEN FUSÉE CHAIN

1. To remove a damaged link, position the rivet over a hole in the staking outfit or a hole in the thirty-six hole vice-stake.

2. Turn a pin punch from carbon steel so that the punch is a little smaller than the pin securing the links; harden and temper it to between brown and purple, then drive the old pin out. If the pin resists strongly, try punching it out from the other side.

3. Prepare the other part of the chain for rejoining.

4. File blue steel to a gentle taper so that it will pass through the link; then tighten it up on one side, which you can, due to the slight taper.

Fig 107 A torque bar for checking the constant pull of a fusée.

5. Cut the blue steel on the thicker side close to the link, then file or stone it so that the end is square and very nearly flush with the link.

6. Using the stake again, tap the pin in so that it is virtually flush with the link.

7. On the other side of the link, cut the pin off close to the link, then file or stone the end square so that it, too, is almost flush with the link.

8. Rest the link on a steel block on the large side of the taper pin and rivet the pin with a flat punch.

9. After riveting, test that the rivet is tight and that the chain is free. If the link is a bit tight, work it to and fro until it becomes free.

11 PIVOTS, JEWELS AND PLATFORM ESCAPEMENTS

Already we have seen that brass holes in plates can wear, pivots can become cut by small, hard pieces of material being lodged in pivot holes, and we are aware of the frictional losses between pivots and pivot holes. These problems can be reduced by the use of jewels which are harder and less porous than brass. Larger clocks can be jewelled, although it is more usual to find jewels on smaller, better quality clocks.

PIVOTS AND JEWELS

Straight Pivots

The endshake of a train wheel, and usually the pallets, is limited by the shoulder of the pivot coming against the plate; in fact most of the friction and wear is caused by the side pressure of the pivot in the hole. This can be reduced by fitting what is called a 'flat jewel hole' into the brass plate, and using the jewel as a bearing.

Cone-Shape Pivots

The balance of a clock cannot work properly with straight pivots because they need even smaller frictional losses, and so sometimes a cone-shape pivot is used, working in what is called a jewelled cup screw. This arrangement also makes the balance relatively shock-proof.

The Conical Pivot

To reduce frictional losses still further on the balance staff, conical pivots are used in conjunction with a cap jewel (which is often called an endstone) and a convex jewel hole. The pivot has a rounded shoulder for strength, though the rounded portion must never enter the hole in the jewel. The convex jewel has two particular features: first, the hole has curved sides – described as 'olived' – thus reducing the friction even further between pivots and hole; second, the top of the jewel is convex, which helps the oil stay around the pivot.

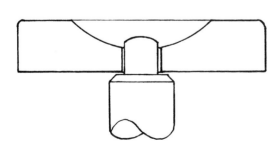

Fig 108 A straight pivot working in a flat jewel hole.

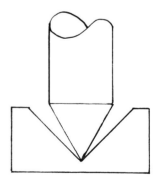

Fig 109 A cone-shaped pivot working in a jewelled cup screw.

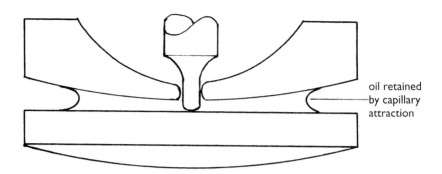

Fig 110 A conical pivot working in a convex jewel hole. Oil is retained by capillary attraction.

oil retained by capillary attraction

Fluids will fill the smallest space available to them, so by having a flat side on the cap jewel facing the convex side of the jewel hole, oil is retained in the middle. At this narrow point, cap jewel and jewel hole should be ³⁄₁₀₀mm apart.

There is a similar arrangement on the other end of the balance staff. If it were drawn, the rounded end of the pivot would be clear of the upper cap jewel by about ³⁄₁₀₀mm or a little more to allow for endshake. Conical pivots can also often be found on the escape wheel and pallet on platform escapements, particularly those fitted to French carriage clocks. This is because the power reaching the escapement is small, and the frictional loss due to the weight of the escape wheel and pallet is considerable. Train wheels can also

have conical pivots with cap jewels and convex jewel holes, though their value is questionable due to the high ratio of power transmitted to weight of the wheels. Most of the friction is sideways.

Pallets

On platform escapements which are the jewelled lever type, instead of pallet pins, jewelled pallet stones are fitted.

Impulse Pins

Impulse pins are sometimes made of steel, but in better quality clocks they are also jewelled. Pins of different shape exist, but by far the most common is the 'D' shape.

Fig 111 Jewelled pallet stone.

Fig 112 'D'-shaped impulse pin.

PLATFORM ESCAPEMENTS

Platform escapements have the significant advantage of being completely portable, and they will function at any angle. Some clocks that were originally designed to work with a pendulum are found with a platform fitted by the manufacturer. Commonly, platforms are found in small, mantel eight-day clocks, carriage clocks, ships' clocks or even chiming clocks, and may be of the club tooth variety, jewelled lever or pin pallet, or they may be ratchet tooth or cylinder. The platform may be fitted to one of the plates with the escape wheel working in the same plane as the rest of the train, although often the escapement is planted across the plates with the escape wheel at right-angles to the wheel which drives it.

To enable the escape wheel to be planted at right-angles to the rest of the train, the wheel that drives the escape wheel has teeth formed at right-angles to the rest of the train and is known as a contrate wheel. This arrangement has implications for the endshake of the contrate wheel. Thus when two meshing wheels are in the same plane, endshake is not critical because a wheel can operate quite satisfactorily anywhere along a pinion; but when a contrate wheel is used, the endshake in it becomes critical because movement will alter the depth between it and the escape wheel.

The solution is to control the endshake by fitting a means of adjustment to one of the plates. The arbor for the contrate wheel is manufactured with a generous endshake. The front pivot works in the front plate in the usual way, but the back plate is provided with a cock and an adjustable screw which is held in the adjusted position by a tight thread. The latter is achieved by slitting the cock and gently closing the threaded hole until the screw can be inserted and adjusted but will hold in the adjusted position. With this arrangement, whilst the shoulder of the front pivot can rest against the inside of the front plate, it is the end of the back pivot touching the screw in the cock that limits the endshake. As will be explained later, it is essential that the contrate wheel endshake is adjusted first, and only then that the platform is fitted and depthed to the contrate wheel.

Fig 113 A contrate wheel.

Fig 114 A cock and screw for adjusting the endshake on the contrate wheel.

Overhauling a Platform Timepiece

After removing the movement from the case, let the power off the mainspring in the recommended way with a winding key or letting down tool. It is possible to let the mainspring down by removing the platform, but there is a significant risk of damaging the contrate wheel teeth and/or breaking the escape wheel lower pivot, so this is not recommended. Should it be necessary, hold the contrate wheel stationary while unscrewing the four platform retaining screws, lift off the platform, then let the clock run down, remembering to interrupt the train periodically to avoid possible damage.

With the platform removed, unscrew the contrate wheel adjusting cock, and repair, clean, reassemble and oil the movement in the usual way. Oil the contrate wheel back pivot as normal, but also put a small drop of oil either on the end of the pivot or on the end of the adjusting screw in the cock.

Replace the cock and adjust the endshake of the contrate wheel to about $\frac{5}{100}$mm; less could lead to pinching because the contrate wheel arbor expands with rises in temperature, too much more could lead to erratic and faulty depthing between contrate wheel and escape pinion.

Dismantling a Platform Escapement (The Jewelled Lever-type)

As dismantling progresses, load the cleaning basket to avoid a cock or bridge damaging a scape wheel or pallet in the cleaning machine. Close examination of the platform is unnecessary at this stage.

1. Remove the one, or sometimes two screws securing the balance cock.
2. Carefully lift the balance cock, keeping it as near as is possible parallel to the bottom plate to avoid breaking a pivot. As soon as it is free, turn the cock which should bring the impulse pin free of the notch in the lever; then lift the whole assembly clear of the platform and place it on the bench. Don't be concerned that the balance is hanging on the end of the balance spring: provided it is not shaken roughly, no damage should occur.
3. Turn the balance and cock over to gain access to the boot. Sometimes the index is fitted with a boot and an index pin, at other times two curb pins are used, and usually the latter when the balance spring has an overcoil rather than a flat balance spring.

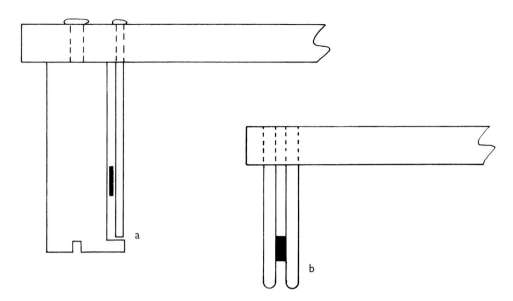

Fig 115 Regulators for flat balance springs (a) and overcoils (b).

4. Turn the boot in either direction by 90° to free the balance spring. If the boot is very tight, rather than shear off the rivet, bend the index pin away from it to free the spring.

5. Pick up the cock and slacken the stud screw by a couple of turns, then push the stud out of the cock. It is usually unnecessary to unpin the balance spring. Tack down the stud screw into the empty hole so that it doesn't come out in the cleaning machine.

NOTE: If the stud is pushed into the cock, use a suitable stake as support close to the stud under the cock, and push it out with a pointed tool; only as a last resort would you unpin the balance spring. If this is necessary, note the pinning point so that the spring can be repinned in the same position; this is important for good timekeeping and helpful in keeping the balance 'in beat'.

6. A large balance should be safe in the cleaning machine, provided its movement is restricted and no heavy parts can knock against it. Alternatively it may be cleaned in the spirit jar with a degreasing agent. For years I have been using lighter fuel because it does the job, is easily available, is relatively cheap and safe to use.

7. Unscrew the two top endpiece screws and place all the parts in the cleaning basket. On very old platforms, the top cap jewel may take the form of what is called a hemisphere, a loose jewel approximating to a half ball: take care with it. These are no longer available in platform size, and finding an alternative can present difficulties.

8. Remove the pallet cock screw, cock and pallets, taking the same care as with the balance cock to avoid breaking a pivot.

9. Remove the escape wheel cock screw, the cock and the escape wheel. Sometimes the pallet and scape wheel are under the same cock, in which case lift them out of their holes vertically to avoid breaking a pivot. If either is stuck with congealed oil, push the pivot out from the other side with the point of your tweezers.

10. Remove the bottom endpiece screw and endpiece, taking care not to bend the two banking pins in the bottom plate of the platform. If present, they may appear bent, but this is probably an adjustment and at this stage should be ignored. Their adjustment is checked with other escapement functions.

Cleaning the Escapement

The escapement is best cleaned in a cleaning machine. The following approximate times will suit most platforms, cleaning machines and fluids:

Cycle	Minutes
Cleaning	2½-4
Spin off	½
First rinse	1½
Spin off	½
Second rinse	1½
Spin off	½
Drying	4 (heater pre-heated but switched off)

Reassembly

After cleaning, parts must no longer be held directly in the hand, but handled in tissue paper.

1. Examine the jewels in the bottom plate for hairline cracks, chips around the hole and congealed oil. The technique for inspection is to use a stronger glass, up to ×10, and to organize things such that the bench light is reflected in the jewels; only under these con-

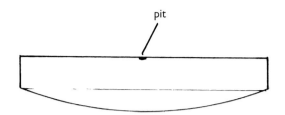

Fig 117 *Inspect a cap jewel for a pit worn by a balance pivot, and for congealed oil.*

ditions will the subtle faults that we are looking for be visible. Cracked and/or chipped jewels should be replaced; jewels with congealed oil need to be cleaned with pegwood so that they are immaculate. Nothing less will give satisfactory results. Remember to peg not only the flat surface of jewels but also the holes from both sides.

NOTE: When inspecting jewel holes, look on the flat side of the jewel and not the countersunk side. A chipped hole is likely to be chipped on the inside edge of the hole, not the outside. Convex jewel holes are inspected on the convex side.

2. Inspect the flat side of the jewel in the bottom endpiece for a pit worn by the end of the balance staff, and for congealed oil. A pit worn in a cap jewel will increase frictional losses, and this will reduce the amplitude of the balance resulting in a loss in timekeeping. A pitted cap jewel should therefore be replaced.

3. Put a drop of light oil on the flat side of the cap jewel in the centre. The oil will form an approximate circle which should cover about half the diameter of the jewel and have a little height.

Fig 116 *Inspecting for hair-line cracks, chips around the hole and congealed oil.*

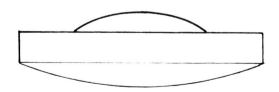

Fig 118 *The cap jewel oiled ready for replacement.*

NOTE: A good oil is likely to have an anti-creep property built into it so that it tends to stay where it is put, and not spread. Also, old jewels are likely to have been treated with an invisible coating to help oil stay where it is put. However, this coating on the jewel is likely to be removed by successive cleaning so we have to rely on the anti-creep property of the oil and capillary attraction.

4. Place the endpiece into its location on the underside of the platform and secure with its screw.
5. From the endpiece side of the platform, inspect for an 'oil circle': this is a circle of oil, looking like a spirit-level bubble, which covers two-thirds of the area of the jewel.

NOTE: When the balance jewel hole is convex, this oil circle usually appears as soon as the endpiece is positioned. If it doesn't, perhaps because the jewel hole and cap jewel have been positioned just a little further apart than the more usual $\frac{3}{100}$mm, feed oil into the hole in the jewel with a pricker until a good oil circle is achieved.

If the jewel hole had a flat surface instead of convex, typically the oil may be drawn over to one side when the cap jewel and jewel hole are brought together. When this happens, keep feeding oil into the hole in the jewel until it builds up

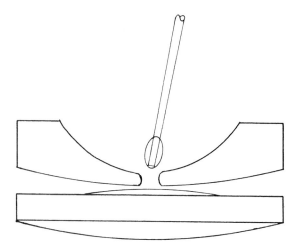

Fig 120 Feeding oil between the jewel hole and cap jewel with a pricker.

as shown in Fig 122; remember that your objective is to lubricate the balance staff pivot. There should be no oil in the countersink of the balance jewel hole.

6. Now examine the escape wheel, particularly the pivots and pinion leaves, with the light reflecting from both in turn. The pivots

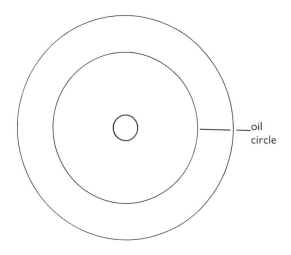

Fig 119 An oil circle covering two-thirds of the area of the jewel.

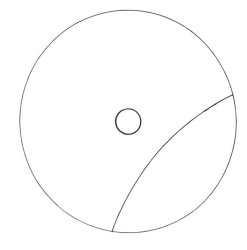

Fig 121 The oil has been drawn over to one side.

should be straight and polished, the pinion leaves should be free of cuts, dirt and rust, and the teeth of the wheel should have sharp edges on the locking corners. Bent pivots should be carefully straightened, and scape wheels with cut pinion leaves should be replaced – although they will have to be retained if replacements are not available, which is more usually the case. Sometimes the contrate wheel can be made to work in a different area of the pinion leaf by fitting washers under the platform, lifting it slightly.

7. Replace the escape wheel in the platform bottom plate.
8. Examine the escape wheel cock jewel hole for hairline cracks, chips and congealed oil. Replace on the platform if in good order, and check the endshake.

NOTE: When manipulating pivots into their holes, use only light pressure on the cock, or a jewel may still chip as a pivot enters a hole.

9. Examine the pallet pivots; check that the pallet stones are secure, and that they are in good order and not pitted and scored by the

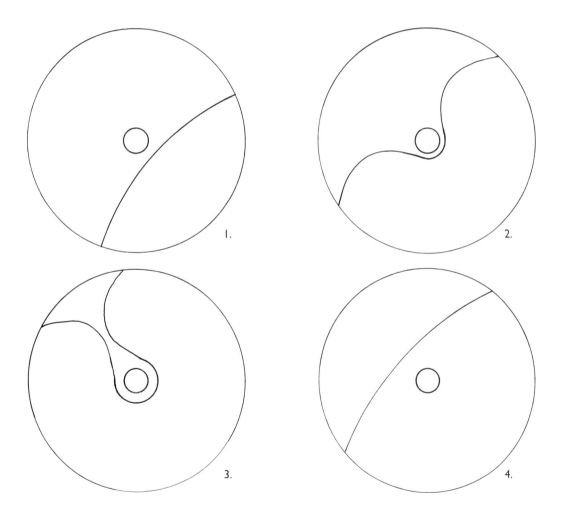

Fig 122 The oil building up when fed through the hole in the jewel. In drawing No 3, the hole still has no oil. When the oil on the left and right of the picture touch, the oil moves to the position in the fourth drawing. The hole is now oiled.

Fig 123 Inspecting the escape wheel. The pivots should be straight and polished, the pinion leaves should have no cuts and the escape wheel teeth should have sharp locking corners.

escape wheel teeth; check that the locking corner and discharging corner of the pallet are not chipped; check the sides of the notch to see that they are not pitted, and check the guard pin to see that it runs parallel with the sides of the notch.

10. Replace the pallets, checking for endshake.
11. Inspect the pallet cock and replace. (Sometimes the escape wheel and pallets share a cock or a bridge.)
12. Oil both top and bottom pivots of the escape wheel and pallets with light oil.
13. Examine the balance cock jewel hole as before.
14. Examine the top endpiece and cap jewel.
15. Oil the top cap jewel on the flat side and place the cap jewel on the bench, with the two screw-holes aligned so that one is towards the back of the bench, the other towards the front. (I am assuming that the endpiece screws enter from under the cock. Some cocks are designed with the cock threaded rather than the endpiece.)

16. Place the regulator in position over the end-piece.
17. Lower the balance cock down over the end-piece so that the screw-hole lines up. Hold the cock in this position without spreading the oil, and replace the two endpiece screws. It is possible to assemble without oil, but I find the difficulty of oiling after assembly a bigger problem.
18. Check for an oil circle as before, adding oil if necessary.
19. If the balance wasn't cleaned in the cleaning machine, wash it now in a small pot with degreasing agent. Dry it on absorbent paper towel.
20. Examine the balance spring for any damage (*see* 'More about Balance Springs' at the end of this chapter). Check the balance pivots for wear and see that they are upright. Check that the impulse pin is clean, upright and secure. Check that there is no split in the safety roller, and that the pipe joining the upper and lower part is not squashed by being driven on too hard by a previous repairer.

NOTE: If the impulse pin is dirty, clean it carefully with pith, Rodico or pegwood. Not every repairer is aware that jewelled impulse pins should never be oiled, and it is common to find congealed oil here which will be harmful if left. A steel impulse pin is lightly lubricated with light oil.

21. Slacken the stud screw to clear the hole for the stud.
22. Place the cock on the bench as it was for

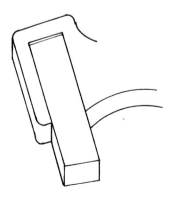

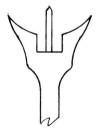

Fig 124 Checking the pallet stones and lever notch.

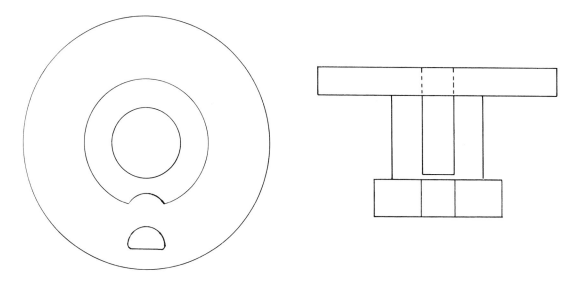

Fig 125 Checking the roller and impulse pin.

reassembly, that is, underside facing up. Hover the balance, with the balance spring facing down, over the cock so that the stud enters its hole and the outside coil of the balance spring lies between the index pin and boot. Assist the stud into the stud hole even to the point where the pin securing the balance spring contacts the underside of the cock.

23. Turn the boot to lock the balance spring between the index pin and boot. As you do this, check that the index pin hasn't accidentally been bent towards the boot to cause pinching of the spring. If it has, bend it out again. The index pin should be upright, leaving a gap of about twice the thickness of the balance spring. If curb pins are used, the balance spring is an overcoil and will need very careful handling. Just enter the overcoil between the curb pins, which must never grip the balance spring but should be in light, permanent contact with it. If this is difficult to achieve, the smallest gap you can cope with is acceptable.

24. Lift the balance cock without turning it over, and lightly tack down the stud screw.

25. Turn the cock over gently, allowing the weight of the balance to be taken by the balance spring. Lower the balance so that the bottom balance pivot enters its jewel hole and the impulse pin enters the notch. Continue to lower the cock, manipulating the top balance pivot into the jewel hole in the cock. Secure the cock with its screw and check the endshake of the balance.

NOTE: The ideal endshake of a balance would be $\frac{3}{100}$mm, or slightly more, but it is essential that the parallel part of the pivot turns against the hole in the jewel. To check for this, put a little side pressure on the balance and move it up and down; there must be no sensation of the end of the pivot touching the olived hole in the balance jewel.

If the endshake is not correct, check that the endpieces have been tightened down properly. If they have and all else is correct, consider removing the balance from the cock, then putting the cock back on the platform on its own and bending (or straightening) it, to adjust the endshake. Don't be tempted to throw up burrs in the cock to correct faulty endshakes; this is often done but deserves no credit. Some tissue paper under the balance cock would be an agreeable solution because it is easily reversed without damage.

26. Having checked that the balance spring is not pinched between index pin and boot, slacken the stud screw and adjust the height of the stud to leave a flat balance spring. The spring itself may need some adjustment for this.

27. With the balance stationary and the impulse pin in the notch, check that the balance spring lies halfway between index pin and boot.

Replacing the Platform and Depthing

It is essential to adjust the endshake of the contrate wheel first, then adjust the whole platform to leave the correct depth between contrate wheel and escape pinion.

1. Having adjusted the endshake of the contrate wheel to about ⁵⁄₁₀₀mm, replace the platform. For your first platform, the theory of gearing to assist with depthing isn't very helpful, so I recommend that you look at the depthing between other wheels and pinions in the clock and adjust the platform to give similar engagement between contrate wheel teeth and escape pinion leaves.

NOTE: When correctly adjusted, there must be clearance between a tooth and two pinion leaves. The clock should 'want' to go after winding by about three teeth of the ratchet wheel. In difficult instances I have even replaced just the bottom plate of the platform with only the escape wheel positioned in the platform and depthed so that the contrate wheel and scape pinion action 'flows': to me, this means smooth, continuous, quiet motion. Also, be careful while adjusting the platform that the contrate wheel teeth don't slip past the escape wheel pinion leaves and become damaged. With the platform replaced and correctly depthed, check the escapement functions (see next section, 'The Club Tooth Lever Escapement'). Next, lubricate the pallet stones and escape wheel teeth.

2. Stop the balance with the exit pallet stone free of the escape wheel teeth. Put a small drop of light oil on its impulse face.

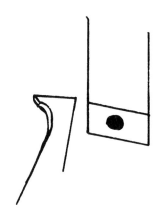

Fig 126 Oiling the pallet stones.

3. Let nearly half the escape wheel pass, then stop the balance with the entry pallet stone free of the escape wheel and put a similar spot of oil on *its* impulse face.

4. Wind the mainspring and check that the balance has a good amplitude. For a lever escapement, a good amplitude would be somewhere between about 240° and just over 300°. When the clock has run for seven days, the amplitude of the balance should still be over 220°. For a cylinder escapement, the amplitude is more likely to be 100° to 160° at best.

5. Test that the balance is in beat: this is when the action of the impulse pin is symmetrical about an imaginary line joining the balance and the pallet centre. Another way of expressing this is to say that when the balance is in the position of rest, the impulse pin should be on the imaginary line joining the balance and pallet centres.

Testing for In Beat

1. Turn the balance by hand in either direction until drop occurs. Immediately it does, stop turning the balance and release it. The clock should continue to go.

2. Turn the balance in the opposite direction until drop occurs; again, immediately it does, stop turning the balance and release it, and

Fig 127 The balance and cock supported on a cock stand for beat setting.

once again the clock should continue to go. If it does go from both sides, it may be said to be in beat; and even if it won't start from either side, it is still (probably) in beat.

3. If it won't go from just one side, remove the balance and balance cock from the platform and turn the balance spring collet on the balance staff to effectively move the impulse pin in the direction in which the balance should have turned when released, but wouldn't.

4. Replace the balance in the platform, and try again until the clock continues to function from both sides. Sometimes the balance cock has a 'beat set device' which means that the stud is secured to an arm that pivots concentricly with the regulator. In this instance, the arm is turned to effect beat setting. When functioning correctly, if the regulator is turned, the beat set device should remain stationary; and when the beat set device is turned, the regulator should turn with it. Often this feature doesn't function properly and it is impractical to correct.

The movement is now virtually finished, so just replace the dial and hands. Case and test the clock.

THE CLUB TOOTH LEVER ESCAPEMENT

The action of the club tooth lever escapement can conveniently be split into three parts to understand the whole: first, the impulse pin and notch action, second, the wheel and pallet action; and third, the safety action.

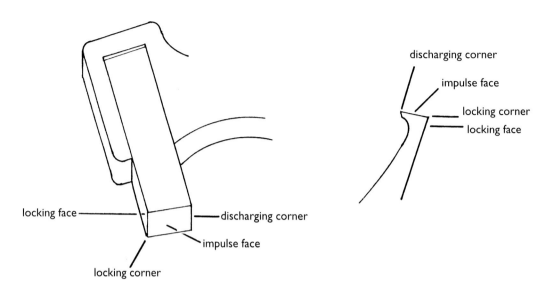

Fig 128 The part names of the interacting parts of the wheel and pallet.

It should be realized that fundamentally the balance should be free to perform according to the natural laws of the balance spring. Of course it cannot be completely free as it needs support, which it gets from the top and bottom jewel holes and cap jewels, and it will need periodic impulse to keep it going. That part of the swing of the balance when the impulse pin is free of the notch is called the supplementary arc and will be our starting point.

The Impulse Pin and Notch Action
The balance comes out of the supplementary arc driven by the balance spring. The impulse pin strikes one side of the notch and immediately starts to push the lever across away from one of the banking pins towards the other banking pin.

Soon the escape wheel is unlocked and gives impulse to the pallet, which means that the opposite side of the notch, under the influence of the impulse, in effect, catches up with the impulse pin and pushes it to begin the impulse to the balance. Immediately the impulse is over, the balance goes into the supplementary arc again. Think of it as similar to an adult pushing a child on a swing: the push is over as the swing leaves

the adult, and the further movement of the swing is the equivalent of the supplementary arc.

The impulse and supplementary arc successively coil up and uncoil the balance spring, with the spring always trying to return to a state of equilibrium leaving the balance in the position of rest, that is, with the impulse pin on an imaginary line joining the balance and pallet centres.

After impulse and before the impulse pin strikes the wrong side of the lever (the outside of

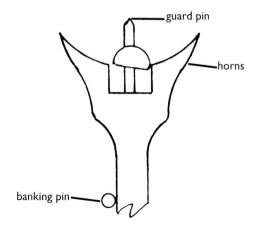

Fig 129 The impulse pin enters the notch.

the notch), the tension in the balance spring drives the balance back again so that once more the impulse pin enters the notch, moves the lever across, then the opposite side of the notch receives impulse from the impulse pin. This process is repeated over and over again. Most older platforms vibrate 18,000 times an hour.

Wheel and Pallet Action

We start with the entrance pallet locked and the balance in the supplementary arc. The impulse pin strikes one side of the notch, moving the lever away from the banking pin. The locking face of the pallet moves away from the escape wheel, sliding across the locking corner of the tooth until both locking corners meet. The unlocking is now complete.

Impulse

Under the influence of the mainspring, the locking corner of the escape wheel slides across the impulse face of the pallet until the locking corner of the tooth meets the discharging corner pallet. Because the tip circle of the escape wheel lies outside a circle passing through the locking corner of the teeth, impulse continues to be given by the impulse face of the wheel sliding across the discharging corner of the pallet. Eventually, both discharging corners meet and the impulse is over. The balance now goes into the supplementary arc.

Fig 131 The unlocking is complete.

Fig 132 The locking corner of the tooth meets the discharging corner of the pallet.

Fig 130 The entry pallet is locked.

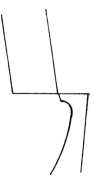

Fig 133 The discharging corner of the tooth meets the discharging corner of the pallet.

It will be noticed that on completion of impulse, the impulse faces of both wheel and pallet form a straight line. It is quite common for this straight line to be shared in the ratio of ⅝ impulse face to the wheel and ⅜ impulse face to the pallet.

Drop

The escape wheel now turns through a small angle of about 1°, which is the drop, then another tooth is arrested by the other pallet stone. This is known as 'first lock'. The locking corner of the escape wheel should contact the locking face of the pallet stone 1½° down the locking face from the locking corner measured from the pallet centre.

Run

The pallet stone is now drawn towards the centre of the escape wheel. The movement of the pallet is called the run, and should be about ½°; it is limited by the lever striking the other banking pin. The run is now complete and we have 'full locking', the lever being held in this position ready for the notch to receive the impulse pin on its return.

This is a complete cycle on one pallet. The whole process is now repeated on the other pallet as the balance returns and enters the notch again. To summarize, the action is: unlocking, impulse, drop, run; unlocking impulse, drop, run.

The Safety Action

Unless provision is made, it is possible that if the clock received a shock, unlocking and impulse could occur at the wrong time and the lever pass to the other banking pin before the impulse pin enters the notch, with the consequence that the impulse pin strikes the outside of the notch and the clock stops; as it surely will. This condition we call 'overbanked'.

To prevent overbanking, a guard pin is fitted under the notch and a safety roller to the balance staff. Under normal operating conditions the guard pin can pass the safety roller through the passing slot. However, in the event of a shock while the balance is in the supplementary arc, the lever moves across, the guard pin strikes the outside of the safety roller, and then the lever is returned to the same banking pin again provided the escape wheel tooth has remained on the locking face of the pallet stone and not come onto the impulse face. The notch, of course, is once again in position to receive the impulse pin when the balance comes out of the supplementary arc.

Testing the Escapement

With the escapement fitted and the balance removed, use tweezers to move the lever away from the banking pin, but keep the escape wheel tooth on the locking face of the pallet. Now release the lever and it should return smartly to the banking pin. Repeat this with the other pallet stone.

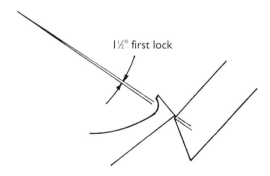

Fig 134 1½° first lock.

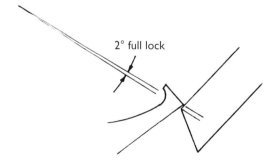

Fig 135 Full lock.

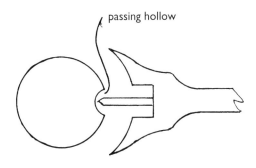

Fig 136 The safety roller with its passing hollow and guard pin.

Lead the lever slowly from banking pin to banking pin and observe the whole action. Make sure that a tooth lands correctly on the locking face of a pallet and that after drop, there is run to banking. With the balance fitted, stop the balance in the supplementary arc and move the lever across until the guard pin touches the safety roller. Still holding the balance, release the lever and it should return smartly back to the banking again. Do this on both pallet stones. Testing the movement of the pallet between the banking pin

and the safety roller is called 'testing the shake on the lever'.

RATCHET TOOTH LEVER ESCAPEMENT

The action of the ratchet tooth lever escapement is very similar to the club tooth lever escapement, the only significant difference being that the tooth has no impulse face so impulse is given only by the tooth sliding across the impulse face of the pallet. The unlocking, drop, first lock, run and full lock remain the same.

THE CYLINDER ESCAPEMENT

Our ideal balance should be free for most of its arc of vibration: such an escapement is classed as a 'free escapement'. The cylinder is classed as a 'frictional rest escapement' because during the supplementary arc the escape wheel is still in contact with the balance. This led to the demise of the cylinder escapement in the early 1900s – it

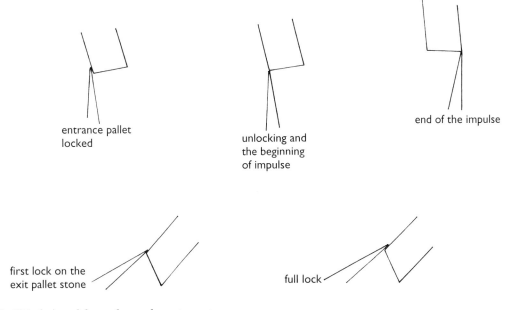

Fig 137 Action of the ratchet tooth escapement.

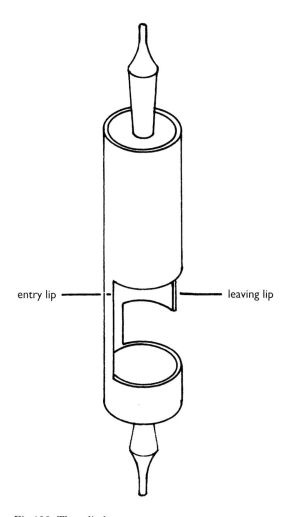

entry lip ———————— ———————— leaving lip

Fig 138 *The cylinder.*

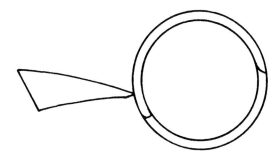

Fig 139 *An escape wheel tooth resting on the outside of the cylinder during the supplementary arc.*

the illustration, the balance is turning clockwise under the influence of the balance spring and the escape wheel is at frictional rest. Eventually the tip of the tooth gives impulse to the entering lip of the cylinder, then the outside edge of the tooth continues to give impulse to the cylinder as the tooth enters the cylinder. Impulse is over when the trailing edge of the tooth passes the entry lip.

Drop
The escape wheel is now free to turn briefly, and this part of the action is called drop. It is the only time when the balance is free of the escape wheel, and it is over when the leading tip of the tooth contacts the inside wall of the cylinder.

was seen as a lost cause in the search for high quality timekeeping.

The escapement has no lever: instead, the escape wheel gives impulse directly to the balance by acting on its arbor which is known as a cylinder. During the supplementary arc, a tooth of the escape wheel rests either on the outside wall of the cylinder, or on the inside wall.

The Action of the Cylinder Escapement
Our starting point is with a tooth of the escape wheel resting on the outside wall of the cylinder and with the balance in the supplementary arc. In

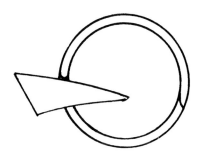

Fig 140 *Impulse being given by the outside face of a tooth to the entry lip of the cylinder.*

Frictional Rest

The cylinder, and thus the balance, will now go into the supplementary arc, of necessity less than 180°. Eventually the balance will be driven back, the 'frictional rest' being complete when the tooth inside the cylinder meets the leaving lip of the cylinder.

Impulse

Impulse is given by the tip of the tooth operating on the leaving lip of the cylinder, and continues once again with the outside face of a tooth giving impulse to the cylinder leaving lip. Impulse is over with the coincidence of the trailing edge of a tooth and the leaving lip.

Drop

The escape wheel is free again until the next tooth drops onto the outside face of the cylinder and the whole cycle starts over again.

The Safety Action

The amplitude of a cylinder balance is limited due to teeth entering the cylinder. To ensure that the balance cannot exceed the maximum safe amplitude, a pin is fitted to the rim of the balance and a second pin is fitted to the balance cock so that should the balance turn too far, for whatever reason, the pin in the balance will touch the pin in the cock.

The escape wheel teeth need freedom inside the cylinder so it is desirable to be able to depth the escape wheel with the cylinder. For this a 'chariot' is fitted under the bottom plate of the cylinder to which the balance cock is screwed. This allows the cylinder to be moved closer to or further away from the escape wheel. Normally there is nothing to be gained by removing the chariot for cleaning and as the depthing is probably correct anyway, it is better to leave it undisturbed.

The chariot has steady pins so if it is necessary to depth the escape wheel and cylinder, slacken the securing screw and push or lever the chariot in the desired direction. There must be freedom or shake of the escape wheel when the tooth is completely inside the cylinder.

Oiling

Lubrication is similar to the procedure for the lever escapement. The cylinder itself needs lubricating with light oil where the escape wheel teeth engage with the cylinder wall.

MORE ABOUT BALANCE SPRINGS

The balance spring does for a balance what gravity does for a pendulum: it is always trying to return to a state of equilibrium, bringing the balance to the position of rest; but in addition, the effective length of the balance can be altered by incorporating a regulator. The effective length of a balance spring is reduced to create a gain and increased to create a loss. From a central position of the regulator where the clock is supposed to keep good time, to an extreme position in either direction will effect a change of between about ten and twenty minutes a day.

Ideally, the centre of gravity of the whole balance, including the spring, should always be in the axis of the pivoting point of the balance; in other words, in the centre of the staff. This is not possible to achieve with a flat balance spring, but to help achieve the ideal, the spring leaves the collet and develops in a spiral getting further from the collet at a uniform rate until about the last half coil. At this point, the spring is bent out slightly to make room for the regulator.

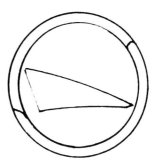

Fig 141 The tooth must be free inside the cylinder.

Fig 142 The flat balance spring with two different outer terminal coils.

Two ways of achieving this are shown in Fig 142: the advantage of the right-hand spring is that the last coil is concentric with the axis of the balance, as is the regulator, so when the regulator is moved, the spring can still vibrate uniformly in the small gap in the regulator. In the left-hand illustration, the balance spring only vibrates uniformly in the regulator when it is central (assuming it has been properly adjusted).

It is also important that the spring leaves the collet in the correct way (Fig 143).

The following checks should be made whenever the balance is replaced:

• The balance spring should be flat from all angles and have the appearance of being reasonably centred around the balance staff and not bunched to one side.

a

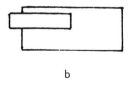

b

Fig 143 The inner terminal coil of a balance spring: (a) shows how the spring should leave the collet in plan view; (b) shows the spring leaving the collet horizontally.

- It must be clear of the bottom of the boot.
- When the balance is turned to an extreme in either direction, the balance spring must be clear of all other parts of the clock and no two coils should touch one another.
- When the balance is in the position of rest, the balance spring should lie half-way between the index pin and boot. This will mean that the balance spring will vibrate uniformly in the regulator even as the amplitude of the balance changes.
- The gap between the impulse pin and boot should be one and a half to two times the balance-spring thickness.

Overcoils

Overcoil balance springs were developed to keep the centre of gravity of the whole balance, including the spring, in the centre of the balance staff. If you look down on an overcoil as a balance vibrates, you should see the flat part of the spring (the body) breathing uniformly around the staff.

There are a number of slightly different overcoils, but by far the most common is the one known as the 'Breguet', which works like this: at about three-quarters of a turn from the stud, the balance spring is lifted, then at a point very close to the lift it is lowered again so that the last three-quarters of a turn is raised above the body of the spring but parallel to it. To achieve this bend, I have an old pair of tweezers with a step filed in one side then, holding the spring at the appropriate point, I push the tweezers into soft wood and effect a lift. Next I would turn the spring over and effect the second bend.

In the plan view of the Breguet overcoil (Fig 146), notice that the arc over which the regulator operates is much smaller than with a flat balance

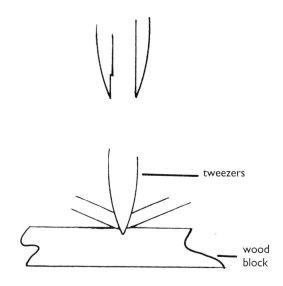

Fig 145 Achieving the overcoil.

spring and that with an overcoil, instead of an index pin and boot, there are two 'curb pins'. When properly adjusted, these should be in light but permanent contact with the spring. Until you develop the skill to achieve this ideal, go for the minimal gap you can cope with.

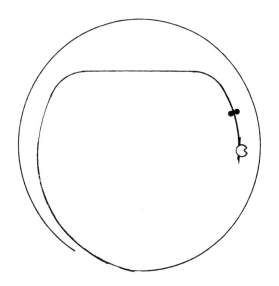

Fig 144 A Breguet overcoil in the making.

Fig 146 A plan view of a Breguet overcoil.

The overcoil is called 'the outer terminal coil' and it can be manipulated to achieve isochronism, a term used when varying arcs of vibration are performed in equal time. For commercial work, adjustment is rarely made and I would advise simply adjusting the overcoil so that it looks symmetrical in plan view, and the body of the spring is centred around the staff.

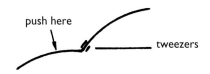

Fig 147 Removing a simple bend in a balance spring.

Most problems with balance springs are self-inflicted: handle balances with care, and you will avoid creating problems.

Correcting Balance Springs
Before leaving the topic of balance springs, a few words about correcting them.

Correcting a Bend in the Flat
With the collet and spring removed from the staff, hold the balance spring on the inside of the bend as close to the bend as you can get. Now manipulate with a suitable tool, nursing the bend out and keeping the tools upright as you work.

Fig 148 Removing a lift in a balance spring.

Removing a Lift
Hold close to the lift with the tweezers, then manipulate 180° away with a suitable tool.

Many corrections are made at the stud. The alternatives include the following:

- Lift or push to achieve a flat balance spring.
- Twist to the right or left to get a balance spring vibrating in the regulator.
- Move the tweezers to the right or left to centralize the spring around the staff.
- Lean the top of the tweezers to the left or right to achieve a flat balance spring.

The above is just an introduction to manipulating balance springs, but forms a basis on which to build skill and knowledge.

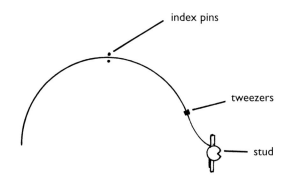

Fig 149 Correcting a balance spring after pinning.

12 FURTHER PLATFORM WORK

There are two common and significant types of repair associated with platforms: one is fitting balance staffs, the other is jewelling.

FITTING BALANCE STAFFS

First we will look more closely at a complete balance to see just what it is. The balance staff is always made of steel, hardened and tempered to dark blue. Referring to Fig 150, 'A' and 'B' are the top and bottom pivots respectively, and 'C' is the diameter which accepts the balance spring collet. 'D' is a steep undercut which is riveted over, securing the balance arms and balance; 'E' is the diameter to which the arms of the balance are fitted. 'F' is the lightly undercut seating for the balance arm, the undercut being to make sure that the balance lies flat and true. This is the hub of the staff, and has what is called a backslope to lighten the staff. 'G' is the diameter which accepts the safety roller.

The collet, balance spring and stud are shown in Fig 151. In the centre is the split brass collet surrounded by the balance spring which may be steel or a modern metal which is rustless and anti-magnetic, and which retains its elasticity with changes in temperature.

The end of the balance spring may be pinned with a brass pin or may be cemented. The balance can be of one metal with a plain rim supported by two or three arms, or it may be a continuous rim with poising screws. The balance may even be split and bimetallic with temperature-compensating screws and timing screws. The roller may be a single table roller or a double safety roller; both carry an impulse pin which is more likely to be jewelled but could be made of steel.

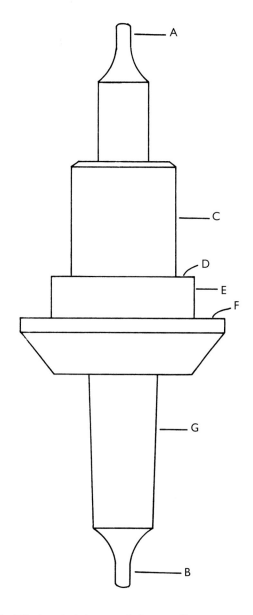

Fig 150 A typical rivet-type balance staff.

Fig 151 Two balances, one a plain balance, the other compensated.

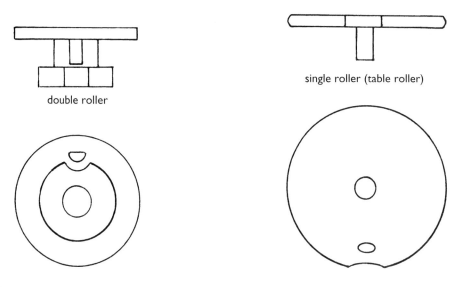

double roller

single roller (table roller)

Fig 152 Rollers.

Dismantling the Balance

1. In order to identify the stud in relation to the rim of the balance, mark the latter with the point of a scribe; such a mark is usually put there by the manufacturer or previous repairer and is known as a 'beat mark'. Note where the impulse pin is in relation to the rim of the balance.

2. Remove the balance spring by inserting a collet-removing tool into the split in the collet, then turning and drawing it off.

3. Hold the balance by gripping the staff on the collet diameter in a pinchuck, then hold the large diameter of the roller in a collet in the lathe.

4. Turn the headstock pulleys by hand, drawing the staff from the roller.

5. Remove the roller from the collet. A dedicated roller-removing tool could be used, but the method described here has given 100 per cent success, whereas other methods have at times resulted in some damage to the roller or balance arms.

6. There are three ways of removing the balance from the staff: two use a lathe, the third involves bursting the rivet using a special tool and a staking outfit.

   First method: Draw the temper of the staff by fitting a brass tube over the roller diameter and applying heat from a spirit lamp. Hold the staff in the lathe by the collet diameter and turn off part of the hub so that the balance comes away, passing over the roller diameter.

   This method holds no real problems. Break the polish on the staff if necessary with an Arkansas stone.

Fig 153 Removing the roller in the lathe.

Fig 154 Drawing the temper of a staff to turn off the seating.

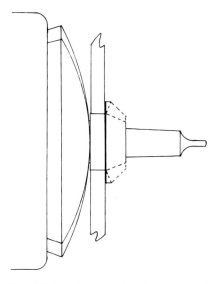

Fig 155 Turning part of the hub and seating away to remove the balance.

Second method: Draw the temper of the staff as above, then turn off the rivet allowing the balance to be lifted off. This method risks a countersink being turned in the arm of the balance, making it more difficult to rivet the new staff. I find this a close second-best to the first method described.

Third method: This method doesn't hold universal favour because on occasions it does involve casualties. However, I feel that when it works it is so quick and easy that it makes up for the occasional mishap.

(a) Select a hole in the staking outfit which is a close fit to the hub of the balance and centre it up. Place the balance staff and balance in the hole.

(b) Position an Unrah-Max tool (available through tool suppliers) over the staff onto the arms, checking that the top pivot of the

Fig 156 Riveting the staff.

balance (or the broken stub) enters the hole in the end of the tool.

(c) Push firmly on the Unrah-Max tool to trap the arm of the balance, then strike the top of the tool sharply with the heel of the palm of the hand. The staff should be driven from the balance.

(d) Remove the tool, and balance and retrieve the old staff to compare with the new.

(e) Inspect the arms of the balance for damage, and to see if the old rivet is still on the hole. It often is and needs picking off with tweezers.

If there is slight curling of the balance arms, they may be flattened by putting the bent arms between a flat stake in the staking outfit and a flat punch, and tapping lightly with a small clock hammer.

7. Compare old and new staffs to make sure they have the same overall height, the same major diameters, and similar, less critical heights.

8. Select a round-ended hollow punch which is a close fit to the collet diameter. Select a hole in the staking outfit which is a close fit to the roller diameter, and centre it up. Clean the staff if necessary.

9. Position the new staff in the centred hole, and place the balance onto the staff. Commonly the balance arms will pass over its diameter and rest nicely on the seating at the top of the hub. If the fit is 'sloppy', the balance may have had an unfortunate history necessitating a new staff to be turned or a new complete balance to be fitted. Only very slight lateral movement of the balance on its diameter is acceptable, as the balance will be centred to a small degree during riveting.

An eccentric balance is not acceptable because first it presents poising difficulties; second, it offends the trained critical eye; and third it is contrary to trade traditions. If the balance doesn't pass over its diameter, turn the diameter of the staff down slightly. Don't be tempted to open the hole in the balance. One of the golden rules in repair work is 'never alter the clock to accept the replacement part, make the part fit the clock'. If the hole in the arms of the balance is an interference fit on the staff, it may be tapped on with a flat hollow punch, the hole of which must be slightly larger than the diameter over which the arms are passing.

10. Once the arms are sitting on the hub, rivet the staff with frequent light blows from a light clock hammer.

11. After a few blows, say eight, remove the staff and balance from the staking outfit, hold the staff by the roller diameter in brass-lined pliers and see if the balance can be turned on the staff. If it can be turned easily, continue to rivet. If it can't be turned or turns reluctantly, it is tight enough.

12. Swap punches now for a flat hollow punch which is a close fit to the collet diameter, and lower the rivet with just a few more light hammer blows.

13. Visually check that the arms of the balance are touching the staff hub all the way round.

Truing the Balance

1. Put the balance in a figure-of-eight truing calliper; spin it, and check that it turns true. It helps to secure a small piece of metal – like a broken drill, for example – with Blu-Tack to one arm of the calliper, adjusting it so that the end of the metal is close to the underside of the rim of the balance. This exaggerates any out-of-true, making it clearer to see. Generally this can be corrected by bending the arms of the balance – with luck, any out-of-true will lie opposite the arms or at the arms. These are the easiest to correct.

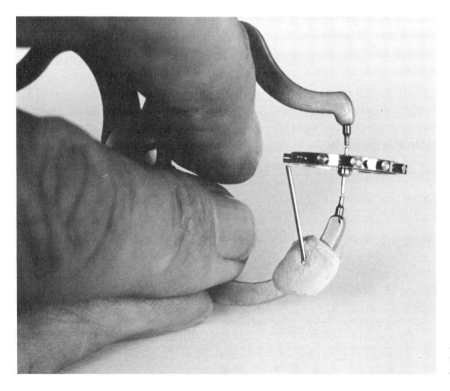

Fig 157 Truing the wheel in a figure-of-eight truing calliper.

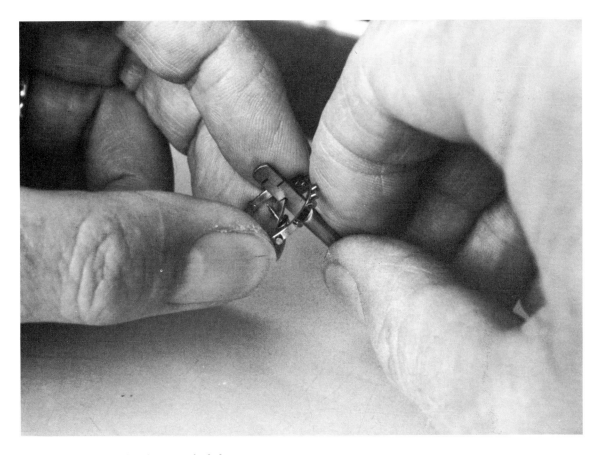

Fig 158 A tool for truing the arms of a balance.

Continue to true the balance until it is perfectly true to the naked eye. As experience is gained, you will be able to correct most balances so that they are true even when inspected with the aid of an eyeglass.

2. Replace the roller. Select a flat hollow punch which is a close fit to the roller diameter.

3. Centre a hole in the staking outfit which will accept the diameter from which the top balance pivot starts, but will not accept the collet diameter, and place the staff in the hole.

4. Position the roller on the staff with the impulse pin pointing in the correct direction. If you are uncertain of the right position, proceed as follows:

(i) Temporarily replace the balance cock on the platform.

(ii) Hold the balance over the cock so that the balance spring stud mark is in line with an imaginary line joining the balance staff and the hole for the stud.

(iii) Position the roller so that the impulse pin points directly to the pallet jewel hole.

5. Drive the roller on so that the large diameter of the roller rests against the underside of the balance seating.

NOTE: If the roller is too tight to go on, turn the roller diameter of the staff down until the roller pushes on to about halfway, then stone and burnish the roller diameter so that the roller pushes on by about two-thirds to three-quarters, leaving the remainder for driving on. A loose roller may have its hole reduced as shown in Fig 160.

Fig 159 Replacing the roller.

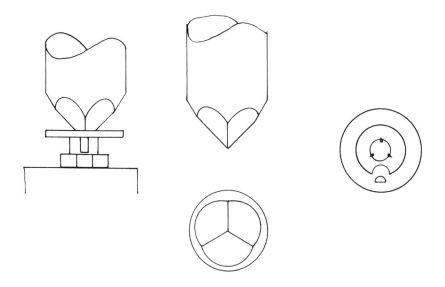

Fig 160 Closing the hole
on a loose roller.

6. Wash the balance, then place it on a poising tool and check for poise.

Checking for Poise

Ideally it will be possible to poise the balance so that it can be stopped in three different positions without any slight heavy spot pulling the balance round. If it does have an obvious heavy spot, there are alternative ways of removing it. For a balance with screws there are three options:

(i) Either, file the head of the screw or screws which fall to the bottom when poising; this would poise the balance but would lighten it marginally, creating a very slight gain, which would, however, be corrected easily by the regulator.

(ii) Or, add a timing washer to the screw opposite the heavy spot; this would have the opposite effect, but is equally easy to correct on the regulator.

(iii) Or, remove metal from the heavy screw and add a timing washer; this would not affect the weight of the balance significantly, and so the rate would not be affected.

For a *plain balance*, drill a light countersink in the rim on the underside. This would be done with a spade drill.

With a *compensated balance*, look for poising screws in the rim of the balance known as quarter screws. In a good quality balance you are likely to find four quarter screws 90° apart; these have relatively tight threads but are not tightened down, and may be adjusted in or out to maintain poise. Try to poise the balance with the minimum of adjustment of any combination of the four screws. If, later, it is found that the clock is gaining or losing by more than the scope of the regulator, use the quarter screws for regulating, adjusting opposite pairs or all four by the same amount to maintain poise. Adjusting

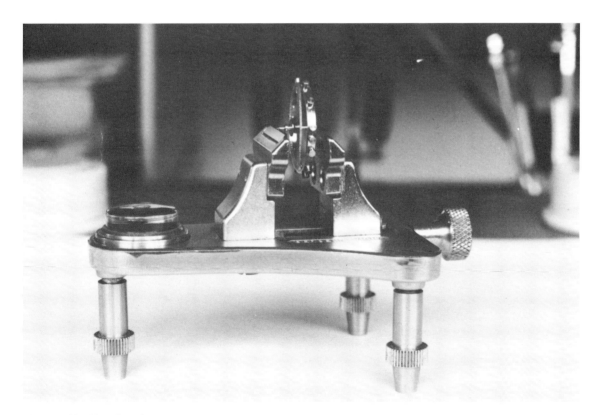

Fig 161 Checking the poise.

them 'in' causes a gain; adjusting them 'out' causes a loss.

Avoid handling these balances directly in your hand when poising because your body temperature can affect one arm of the balance more than the other, creating the illusion of out-of-poise. Handle them with brass tweezers to prevent this happening and to avoid marking them.

These processes are known as 'static poising', and this is all that is necessary for ordinary domestic clocks.

Replacing the Balance Spring

1. Select a flat, hollow punch which is a close fit to the collet diameter.
2. Centre a hole in the staking outfit which will accept the bit of the roller diameter protruding from the roller.
3. Position the balance, and place the collet on the staff so that the stud is in line with the beat mark on the rim of the balance.
4. With the punch already selected, push the collet on so that it rests on the rivet, securing the staff to the balance.
5. Replace the balance in the cock and fit it to the platform. Check the endshake of the new balance, the fit of the pivots in the jewel holes, and that the balance clears the pallet cock, boot and stud. Check for 'in beat' and correct if necessary.

JEWELLING

Types and Sizes

There are five types of jewel used in clock repair work. Modern jewel holes and cap jewels are available in packets of three with their diameters sized in hundredths of millimetres rising in ten hundredth steps. The sizing does not accord with old jewels, which are quite unpredictable, and their shape is different too, which often presents a challenge when a new jewel has to be fitted, as will be seen.

Cap Jewels

These are sized by diameter only, rising in $^{10}/_{100}$mm steps from $^{70}/_{100}$mm to $^{230}/_{100}$mm, though some sizes are omitted in the bigger diameters; for example $^{170}/_{100}$mm and $^{190}/_{100}$mm are not included in the range of Seitz cap jewels.

There is a special type of cap jewel called a 'hemisphere' which is a loose jewel used in the upper endpiece of platforms. Large sizes for platforms have not been available for some years, however, so no further information is given here except to say that an ordinary cap jewel can be used to replace a hemisphere.

Convex Jewel Holes

Sizing is by hole, rising in $^1/_{100}$mm steps from $^7/_{100}$mm to $^{16}/_{100}$mm, and diameter starting at

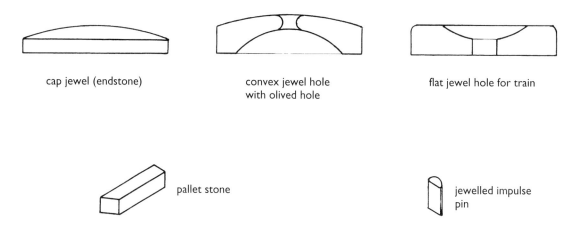

cap jewel (endstone)

convex jewel hole with olived hole

flat jewel hole for train

pallet stone

jewelled impulse pin

Fig 162 Five types of jewel used in repair work.

$^{70}/_{100}$mm rising in $^{10}/_{100}$mm steps to $^{180}/_{100}$mm, although the $^{170}/_{100}$mm is excluded from the range. The main use of convex jewels is for jewelling the balance, though train wheels and pallets do occasionally have convex jewels.

Flat Jewel Holes

Flat jewel holes are sized by hole and diameter similar to convex jewel holes, though the range is bigger. Hole sizes rise in $^{1}/_{100}$mm steps between $^{10}/_{100}$mm and $^{20}/_{100}$mm, then increase in $^{2}/_{100}$mm steps between $^{20}/_{100}$mm and $^{90}/_{100}$mm.

Diameters start at $^{80}/_{100}$mm and rise in $^{10}/_{100}$mm steps in the smaller sizes but with gaps in the range. The largest diameter generally available is $^{300}/_{100}$mm diameter – that is, 3mm in diameter.

Pallet Stones

Pallet stones are selected according to whether they are 'entry' or 'exit', and are sized by their width, rising in $^{2}/_{100}$mm steps from 0.22mm to 0.44mm.

Jewelled Impulse Pins

Jewelled impulse pins are available by diameter rising in $^{2}/_{100}$mm steps from $^{32}/_{100}$mm to $^{54}/_{100}$mm.

Fitting Jewels

There are two basic ways of securing cap jewels, balance jewels and train jewels. The early method, which can offer so many challenges to us, is known as 'wipe in', where the brass plate is wiped over to secure the jewel which is an interference fit in the hole. A variation on this was to fit the jewel to a bush and then fit the bush to the plate: this is known as a mounted jewel.

The old original jewel will probably not have a nice convenient outside diameter, but will be an odd size, and the difficulties arising from this can be threefold: first if the replacement jewel is smaller than the original in diameter, can the concentricity be maintained? Second, how can the hole in the brass be stretched to accept an oversize jewel yet maintain concentricity? And last, if the replacement jewel is slightly higher than the original, can it still be wiped in?

The alternative to 'wipe-in' is 'friction-fitting', where the hole in the plate is opened to $^{1}/_{100}$mm under the size of the jewel diameter. The hole is produced by a parallel reamer, and the jewel is pushed into the hole and held by friction. This is similar to the way a bush is fitted, except that a jewel hole need not be flush with the inside of the plate as a bush is.

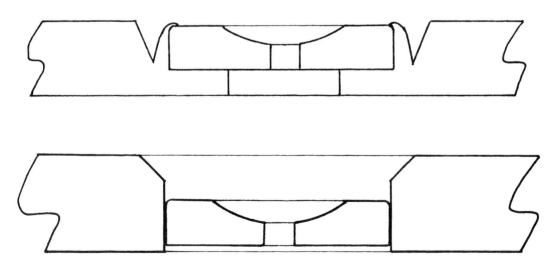

Fig 163 (upper) A wipe-in jewel.
Fig 164 Friction-fitted jewel.

Tools at our Disposal

A jewelling press is shown in Fig 165 (a Seitz), accessories in Fig 166. At the top of the tool, immediately under the handle, is a micrometer scale enabling jewels to be positioned with precision.

Replacing a Wipe-in Jewel

1. Push out the old jewel using a flat pusher with a suitable stake for support; the stake should be positioned close to the hole. Adjust the micrometer so that the tapered part of the pusher doesn't damage the setting.

2. If it is complete, measure the diameter of the old jewel. If it is broken up, assess the diameter by opening the 'wiped in' part of the setting almost fully and trying a new jewel into the hole.

3. Determine the hole size for the new jewel by trying the pivot into one of the jewels in the pivot gauge supplied with the jewelling outfit. A good fit will allow the wheel and pallet arbor to lean by about 5° in all directions.

4. Select a new jewel from stock and place it over the pivot to confirm that the size is correct.

5. Now, if necessary, open the hole in the plate a little further. Ideally the new jewel should be an interference fit into a fully opened hole. Expect to compromise, but work towards keeping the jewel concentric with the original.

6. Assuming that the jewel is an interference fit into the hole and that it is lower than the wall, proceed to 'wipe it in' with a suitable, lubricated tool. Use light pressure at first, then gradually increasing. Finally, test that the jewel is secure by gently pushing on its other side with pegwood.

Fig 165 Seitz jewelling press.

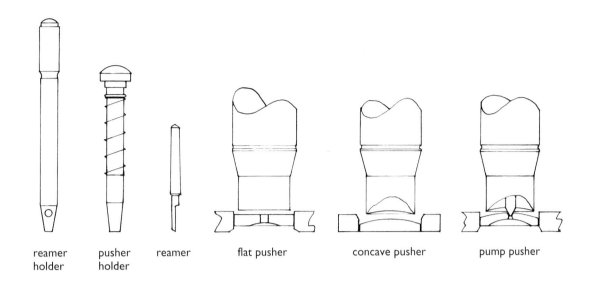

reamer   pusher   reamer   flat pusher   concave pusher   pump pusher
holder   holder

Fig 166 Accessories for the Seitz jewelling tool.

If fitting a jewel by 'wiping in' is unsuccessful, converting the hole to a friction fit is an option (see next section), although it is not a straightforward procedure: only rarely can the plate be opened to accept a friction-fitted jewel, as there is often insufficient brass behind the 'wipe-in' section to hold the jewel. Furthermore, there is a significant risk of the hole drifting off-centre. To broach beyond the 'wipe in' would certainly mean that the jewel could be held securely, but there is still the risk of drifting off-centre, and the completed job would look ugly with such an oversize jewel.

Even the experienced worker approaches replacing a wipe-in jewel with a certain apprehension, so why not practise on an old platform before tackling one that matters?

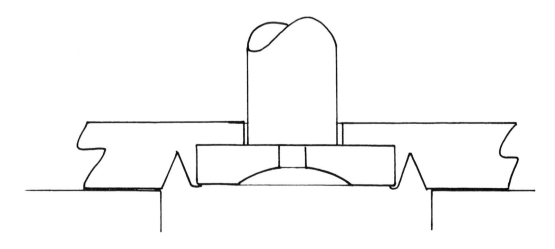

Fig 167 Pushing the old jewel out.

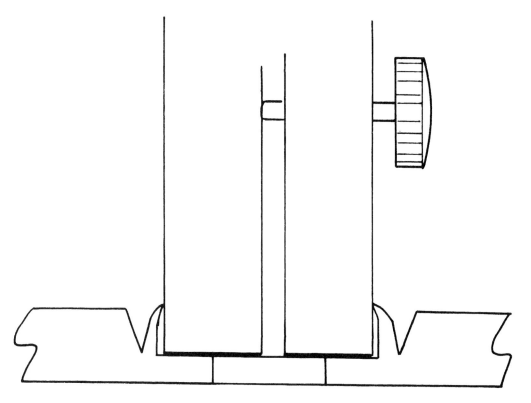

Fig 168 Opening the setting with an opening tool.

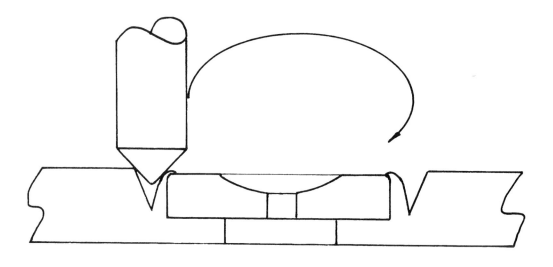

Fig 169 Wiping in the new jewel.

Replacing a Friction Jewel

1. After checking that the pivot is in good condition, determine the hole size for the new jewel using the pivot gauge.
2. Select a suitable stake to use as support around the hole of the jewel to be replaced, and put it in the jewelling press.
3. Select a suitable pusher to remove the old jewel.
4. Position the part to be jewelled on the stake under the pusher, lower the jewel-press handle gently, and adjust the micrometer scale so that the end of the pusher touches the inside of the jewel but so that the jewel could not be pushed out even under pressure. Take a reading from the micrometer.
5. Now adjust the micrometer to allow the jewel to be pushed out. Usually one turn clockwise on the micrometer is sufficient.
6. Push the old jewel out.
7. If possible, measure the outside diameter of the jewel and select a new jewel of the right hole size and outside diameter. If the diameter of the old jewel is an odd size, open the hole with a reamer, making sure the part being broached is held horizontal to prevent an oversize hole being produced. The broached hole may need deburring.
8. Return the micrometer to the original setting, position the part to be jewelled, and push the jewel in to the pre-set level.
9. Finally, check your work. For example, if it was an escape wheel that was having a replacement jewel, on completion check that the escape wheel has the right endshake. Also check the action between the escape wheel and pallet if replacing scape or pallet jewels.

When replacing cap jewels, invariably the flat side of the jewel is flush with the inside of the endpiece and it is important to be aware that it is very easy to damage parts of a platform jewelling. If what you expect doesn't seem to be happening, check on what you are doing before pressing harder on the jewelling press. A common mistake is inadvertently to press on brass rather than the jewel.

SETTING PALLET STONES

Pallet stones are held in position by shellac from the underside. They should be checked when overhauling a platform, to make sure that they are secure and that working faces and corners are in good condition. Before attempting to replace a pallet stone you will need to make yourself an aid: an empty bottom plate from an old Geneva pocket watch makes an ideal tool for securing pallet stones, as it already has a variety of holes. A handle of brass and clock pegwood is a quick and easy arrangement.

1. Select a new pallet stone of the correct width, entry or exit.
2. If the old damaged pallet stone is available, bring the two impulse faces together. If the two pallets form a straight line, the impulse face is the same angle on both pallet stones.
3. Degrease both the pallet and pallet stone in the spirit jar and clean off any old shellac from the pallet frame.
4. Smear a little grease across the plate that you are going to use to prevent the shellac from sticking to the plate.
5. Position the pallet on the plate so that the upper pivot enters one of the holes in the plate and the pallet frame is in full contact with the plate.

Fig 170 Invariably a cap jewel is flush with the inside of the endpiece.

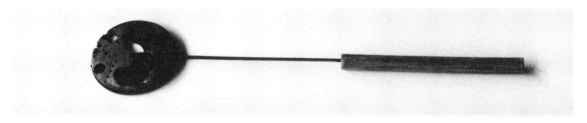

Fig 171 An aid to setting pallet stones.

6. Carefully position the new, clean pallet stone into the frame, placing it in what is felt to be the right final position. Make sure both pallet frame and pallet stone are held flat. Often I hold the pallet staff and frame down on the plate with one pair of tweezers, positioning the pallet stone with a second pair. I also often push the pallet stone into the frame with pegwood to avoid chipping the stone.

7. Chip off a small lump of shellac and break it up into a number of pieces so that two suitable pieces can be selected.

8. Position one small piece of shellac over the pallet and frame and a much bigger piece close to the pallet but so that the pallet is between it and the flame. The purpose of the second piece of shellac is that it is seen more easily than the smaller piece on the stone and allows a better judgement to be made on the melting of the shellac.

9. Hold the watch plate over a spirit lamp and allow the heat to travel to the pallets. Classically, the larger piece of shellac will start to change shape, forming a round, shiny droplet and then spreading a little. The next phase is that the shellac will start to bubble, but try to remove the watch plate from the flame just before it happens.

NOTE: Care must be exercised to avoid overheating the pallets. Avoid bluing or even colouring the pallet frame or pallet staff.

10. Allow the plate to cool sufficiently to remove the pallet from the plate without disturbing the stone from the pallet frame.

11. Degrease the pallet then try it in the platform with the escape wheel and put power on to try the escapement functions. In particular, make sure that you have drop, safe lock and run to the banking.

It may be necessary to warm the pallets again and adjust the pallet 'in' or 'out' to achieve a correctly functioning escapement. If the new stone was an interference fit in the pallet frame, as it ought to be, changes to the angle of draw should be unnecessary.

REPLACING AN IMPULSE PIN

The process of securing an impulse pin is very similar to securing pallet stones. With any luck, the old pallet stone will simply be loose, but if it is missing altogether, a new stone of the same size will have to be found.

1. Remove the roller with a roller removing tool, or in the way described on page 127.

2. Clean off the old shellac by 'digging' at it with

Fig 172 Testing that the replacement pallet stone has the same angle as the original.

*Fig 173 Ready to shellac
the stone into place.*

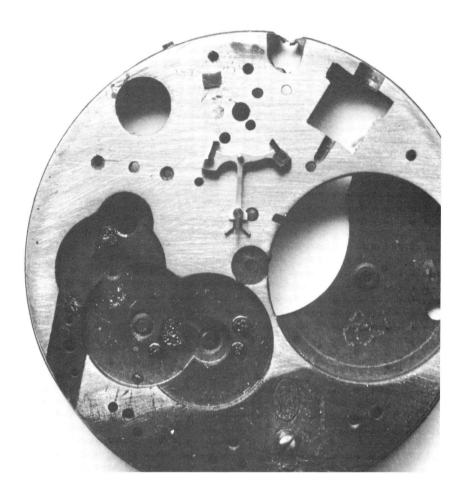

pegwood or with a fine brass scraper which
will not scratch the roller.

3. Most workshops seem to have an assortment
of impulse pins, so select one of suitable
length which is an interference fit into the
roller. This is a tricky job needing good
tweezers. Before proceeding, check that the
impulse pin fits the notch in the lever with
perceptible play: there must be play!

4. Degrease both impulse pin and roller and fit
the impulse pin into the roller, checking that
it is upright.

5. Place the roller on the watch plate as recommended for pallet stones, positioning a small
flake of shellac on top of the impulse pin.

6. Warm the plate over a spirit lamp, and watch
for the shellac melting and flowing around

the impulse pin.

7. Remove the plate from the flame, allowing it
to cool and the shellac to harden off.

8. Check that no shellac has found its way too
close to the hole in the roller or it will crack
when the roller is driven back onto the staff.

9. Check that the impulse pin is upright and
secure. If it is, reassemble the balance. There
should be no need to poise the balance after
this under normal circumstances.

If the impulse pin is out of upright, feed the
roller over a length of tapered brass pin-wire,
then bring the roller close to the heat of a hot
soldering iron. This will soften the shellac
allowing adjustments to the impulse pin to be
made. This may sound unrefined and even
unprofessional, but it is quick and effective.

13 PART JOBS

RE-WAXING, SILVERING AND LACQUERING AN OLD DIAL

When you see a properly wax filled and silvered dial it looks very impressive and is likely to be admired by anyone who sees it. The successful dial restoration oozes skill, knowledge and specialist technique – but in fact all that is needed are one or two basic tools, easy-to-obtain materials and a little care.

Let us assume that you have a flat, circular, waxed-filled, silvered dial that has lost some of its wax, and that the silver plating has worn off exposing some of the brass underneath. Before starting to restore it you will need to purchase a silvering kit, which may include a stick of hard – it must be hard – black wax and some clear lacquer. These can be obtained from any materials house. You also need a tool to remove the surplus wax, one sheet of 350 grade wet-and-dry, a board for circular graining of the dial, and heat such as a Calor gas torch.

The tool for removing the worst of the surplus wax may be made from a brass strip 50 × 25 ×

Fig 174 A tool for scraping off the surplus wax, together with a graining board.

3mm ($2 \times 1 \times \frac{1}{8}$in) thick fitted to a file handle. A board for graining can be made from any flat board of about 45cm (18in) square and about 2cm ($\frac{3}{4}$in) thick; in the middle fix an upright dowel protruding from the board by 2cm ($\frac{3}{4}$in). Make up an arm $45 \times 5 \times 2\frac{1}{2}$cm ($18 \times 2 \times 1$in) with a hole at one end a close fit to the dowel in the board, and with another dowel fitted in the other protruding by about 10cm (4in). Wrap a strip of 350 grade wet-and dry around the arm, and pin with a couple of drawing pins.

To use (after waxing), position the dial centrally, then hold the centre of the arm down with one hand and swing the arm round in a circle with the other hand until the dial is uniformly grained. Double-sided tape can stick down a flat dial; dials with feet will probably mean that holes need to be drilled in the board to accommodate them.

Removing the Old Wax

1. Make a suitable tool for removing the old wax; a thin and a wide scraper similar to a scribe and a screwdriver will be needed, particularly if the numerals are Roman. Scrape out all the old wax leaving, if possible, bright brass to which the new wax can adhere. Be careful not to scratch the silvered part of the dial.

2. Wash the dial with soap and water, rinse and dry.

3. Place the dial on the lid of a biscuit tin and warm it with a flame, all over or locally, until the warmed dial melts the hard wax when it is brought to the dial. Don't put the flame on the wax, and avoid getting the dial too hot. Work your way around the dial filling all the engraving, which will include the chapter ring and the name on the dial if present.

4. When the wax is set but the dial is still quite warm, heat the brass scraper and pull it across the dial, removing the worst of the surplus wax and tending to push the wax down into the engraving.

5. Position the dial centrally on the graining board and remove all the surplus wax from it, leaving clean, filled engraving and a clean bright surface with a uniform concentric grain. This could also be done on a centre

lathe with the dial secured to a piece of wood which is itself fastened to a faceplate.

6. Remove the dial and wash it with soap and water, rinse and dry. Avoid touching its front which is to be silvered.

7. Apply the silvering paste in accordance with the instructions with the silvering kit.

Silvering

1. Starting with a clean dial, sprinkle about half a teaspoon of silvering powder (Horosilv) onto a clean clock glass.

2. Dampen a wad of cotton wool with cold water, dab it on the silvering powder, then rub the cotton wool on the dial, working locally and moving the cotton wool in small circles. Work your way across the whole dial gradually until a uniform silvered finish is achieved. Recharge the cotton wool as necessary.

3. Rinse the dial in cold water.

4. Sprinkle a lesser quantity of Horofinish powder onto another piece of glass, dampen another wad of cotton wool, dab it onto the powder and work in circles across the dial again.

5. Rinse off for the last time in cold water and dry thoroughly with a soft, clean cloth.

Lacquering

Lacquer for use with silver may be painted on with a broad lacquering brush, but personally I prefer to lacquer with clear lacquer spray. On occasions I have experienced a small but unacceptable running of wax after spraying, and the solution has been to spray a thin coat of lacquer initially to seal the dial, then to add a further heavier coat.

Allow the lacquer to harden right through for twenty-four hours before handling the dial.

I make up my own silvering compound, but I would not recommend doing this as the process can be dangerous and the end result is no better than a proprietary silvering kit.

WARNING: **Whether you mix your own silvering compound or use a proprietary brand, all silvering compounds contain silver chloride** and are poisonous by ingestion. Use as directed by the manufacturer, taking additional sensible precautions such as cleaning down after use, washing your hands, and storing the kit safely.

SPLICING A LONG-CASE CLOCK ROPE

One of the many bonuses of training adults is that they often bring specialities with them and so the roles of student and teacher are briefly reversed to mutual benefit. Few repairers seem to know how to join a twelve-strand long-case clock rope competently (which included me), so I was pleased to learn from one of my students and to pass on here what makes for an efficient, good-looking join.

To do the job have the following ready: a rope of the correct length, scissors, masking tape, old general-purpose tweezers bent in the middle so that there is a large gap behind the points, a marline-spike or similar (I use an oval burnisher ground at the point to resemble a marline-spike),

a pen and whipping twine (crochet cotton was an excellent substitute for whipping twine).

The secret of successful splicing is in the preparation, so take the time to get it right and the result should be a good splice that is strong and will pass through the pulley and great wheel.

1. Start by unpicking a short length at the end of the rope to see how many strands you have, and how many strands are in the core. Here the rope has twelve strands and the core four.
2. Wrap masking tape around the rope 100mm (4in) from each end to prevent the rope unravelling more than you need, but not too tight, because the strands have to slide over the core later.
3. When you have unpicked about 50mm (2in), secure the ends of each of the twelve strands in the rope with small pieces of masking tape. This is a tedious but very necessary part of the job. Do this at both ends of the rope, then finish unpicking the first 100mm (4in) of each end.

Fig 175 The ends have been prepared.

Fig 176 Sliding the rope
back over the core strands.

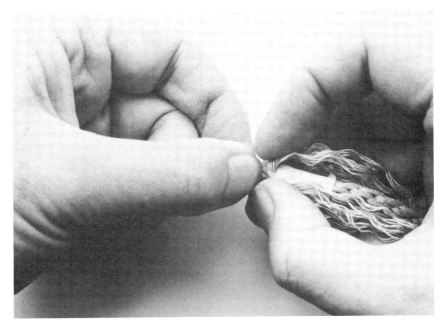

Fig 177 West Country
whipping.

4. Hold the four core strands with your finger-tips close to where you wrapped masking tape around the whole rope, and slide the rope back over the core strands by about 100mm (4in).

5. Cut the core strands at the point you are holding. Do this at both ends of the rope.

6. Select any two core strands from each end and shorten them by a further 50mm (2in).

The next operation is joining the cut ends, and for this you will need to learn West Country whipping. It's quite simple: just practise wrapping a piece of string around any object of about the same diameter as your rope, tie a simple knot, then turn the object over, bring the two ends of the string round and tie another knot. Repeat this a number of times, then finish off with a reef knot. Now for the real thing:

Fig 178 The butted ends secured with West Country whipping.

7. Lay a piece of fine whipping twine on a flat surface, positioning your rope over the twine so that the cut ends of the core of the rope butt together. That is, a long end from the first end and a short end from the second end must butt; and the short end from the first rope and the long end from the second rope must butt.

8. Now secure with four lots of West Country whipping.

9. Slacken the masking tape if necessary, then slide the rope back over the core so that the two ends butt together.

10. Tie two lengths of whipping twine around each end of the rope where they butt to reinforce the job of the masking tape.

11. Now pair off the twelve strands at each end in turn so that each end will have six strands instead of twelve. Pick strands close together. Tape the pairs together, then mark the six ends from one end of the rope for easy identification.

12. Place the rope in front of you with the rope lying left to right. Now pass a double strand from the left to the right, then a double strand from the right to the left. Keep this up for all six double strands.

Fig 179 Six pairs of strands from each side passed to the other side and tied. The arrow points to the butt.

13. Tie off one set of six strands to keep them out of the way, say the unmarked six.

14. Remove the masking tape from the untied section of the rope where the marked strands overlap and where the splice will begin. Leave the whipping twine for the moment to enable a neater splice to be achieved.

15. Take your marline-spike and enter it under two adjacent strands of the standing part of the rope to lift the strands.

16. Remove the marline-spike and push the tweezers through the same gap.

17. Now take the nearest marked free end and pass it through the tweezers.

Fig 180 Marline-spike under two strands for the first tuck.

Fig 181 The nearest marked end passes through the tweezers.

18. Pull the tweezers out, pulling the marked strand under the two lifted strands.

19. Take the next marked strand and do likewise, but choosing the next convenient pair of strands to pass under.

20. Do the same with all six strands and you will have taken your first tuck on one side.

21. Now do exactly the same with the other side after removing the whipping twine holding down the other six pairs of loose ends and the masking tape. For the moment, keep in place the whipping twine that you tied at the head of the masking tape.

22. As work progresses have a tug at all loose ends to avoid an over-loose splice. The first tuck on both sides is shown in Fig 182. For the second tuck on both sets of six pairs, the principle is to go over two strands, then under the next two strands.

23. Now take one further tuck on both sides with all ends. At this stage the two remaining pieces of whipping twine that were at the head of the masking tape can be removed. After your first tuck they serve no purpose.

24. With two tucks in both ends completed, place the splice so far done in the palm of your hand, place the other palm on top and roll the splice several times under pressure. It should be starting to look tidy.

25. So far we have been working with six pairs of strands at each end of the splice. Now it is time to separate each pair in turn and to cut out one strand of each pair. This is the beginning of tapering down to a finish. Have a light pull on the rope, then cut out the strands as indicated but not too close to the rope; for the moment this will leave 'feather' ends which will be cut back later.

26. Working on each end of the splice in turn, take a further two tucks, going over one strand, then under two.

27. The final stage is to 'parcel' the strands: simply push the tweezers up through the centre of the rope, about 2 to 3cm (¾ to 1in) from the loose ends, pull the loose ends down through the centre, and then having stretched the rope, cut off the loose ends and any remaining 'feathers'.

This splice took my bodyweight of 72kg (160lb) without any visible signs of parting. The average long-case weight is only about 5kg (11lb).

Fig 182 The first tuck is taken on all strands on both sides.

Fig 183 The finished splice.

CLOCK MAINSPRINGS

Traditionally mainsprings were made of steel, hardened and tempered to dark blue. Modern mainsprings are more generally white or brown in colour, although they have similar characteristics to the old blue mainsprings; however, doubts are often expressed about the comparative quality of modern mainsprings to the extent that some horologists advocate the continued use of an old spring in a clock even though it may be past its prime.

Mainspring failure is usually because:

(a) It has broken somewhere along its length.
(b) Its outer end is torn and therefore slips round the wall of the barrel.
(c) It has become 'set': this means that it is fatigued, and when removed from the clock, remains fairly close-coiled rather than open-coiled.

Broken Mainsprings

When a mainspring breaks near its inner end, it does so when it is near its fully wound state. The result is a violent kick of the barrel which often bends or breaks its teeth and can damage the intermediate wheel teeth, arbor and pinion leaves. A mainspring that breaks at, or near its inner end must be replaced. However, assuming that the original spring was of the correct length, a repair to its outer end is quite possible, provided the length is reduced by no more than 10 per cent. The procedure for repairing the end is similar to that for repairing a torn end.

Mainsprings are more usually measured in metric units and can be sized by height (H), force or thickness (F) and length (L) in that order. When ordering a new mainspring, state which type of end is required: for example when the mainspring is fitted into a barrel it has an eye end;

loop end

eye end

riveted loop

Fig 184 Three different mainspring ends.

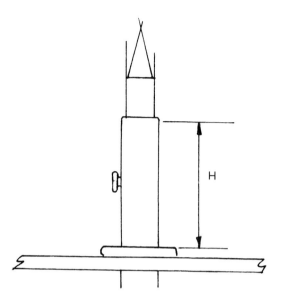
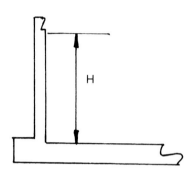

Fig 185 Determining mainspring height.

an American strike has a riveted loop end; and a small alarm often has a loop end.

How to determine the height of a mainspring is shown in Fig 185.

The force of a mainspring that fits into a barrel can be determined by the inside diameter of the barrel, the rule of thumb being that the mainspring force should be $\frac{1}{100}$th of the inside diameter of the barrel. In other words, if the inside diameter of a barrel were 45mm, then the force of the mainspring is likely to be 0.45mm. When ordering a new mainspring it would only be necessary to state its height, the inside diameter of the barrel and the type of end; the length should be right automatically. Other rules of thumb for mainspring, barrel and arbor are:

(i) The space inside a barrel should be occupied by one-third mainspring, one-third space and one-third arbor.

(ii) The mainspring should be about one-thirtieth to one thirty-second of the barrel arbor diameter.

Repairing a Torn or Broken Outer Eye End
As already mentioned, the outer end of a mainspring can be repaired; springs are sometimes too long anyway, and the clock will actually work for longer on a shorter spring of the correct length.

1. Using shears, cut off the broken outer end of the mainspring close to the old eye, and file to shape.
2. In a gas flame, soften the end of the mainspring by about 3cm (1in) by bringing the spring to a cherry red and allowing it to cool slowly.
3. Mark the required shape for the new eye on the mainspring, making sure that it is large enough to pass over the hook in the barrel wall.
4. Select a flat punch that will just enter a hole in the staking outfit that is a little smaller than the eye, and centre the hole.
5. Enter the mainspring between the punch and hole, and drive out a circle of metal to accept a needle file. Clear the hole in the stake.
6. Finish the eye using a needle file.

Fig 186 The marked eye.

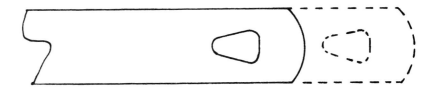

Fig 187 The centred hole in the stake (below).

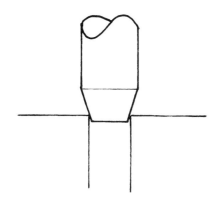

After this, the end of the mainspring is not hardened but left in its soft state. Do not be tempted to repair the inner end of a mainspring as it is unlikely to be successful, and there is significant risk of personal injury.

Set Mainsprings
There is no effective, lasting way to recover a tired, set mainspring. If, when a mainspring is removed from a barrel, it doesn't open up by more than about twice the diameter of the barrel, replace it with a new one. Exceptions to this would include a very old mainspring of interest to conservationists.

Replacing a Barrel Hook
1. Select a piece of silver steel (or mild steel) which will pass through the eye in the mainspring with clearance.
2. Drill a hole in the barrel wall which will be halfway up the inside wall of the barrel, and half the diameter of the selected steel. (Beware: halfway up the inside wall will

probably not be halfway up or down the outside wall.)
3. Deburr and lightly countersink the outside of the hole by scraping. Do not use a countersink tool. Due to the curvature of the barrel a countersink drill will not countersink uniformly around the hole.
4. Set up the selected steel in the lathe so it has minimal overhang, and turn a slightly tapered diameter with a square shoulder so that it enters the drilled hole. The length should be just greater than the barrel wall thickness. Hollow the end for riveting.
5. Turn a further step, the length of which should be about $^{10}\!/_{100}$mm greater than the spring thickness. This time, taper the shoulder to the proportions shown in Fig 189.

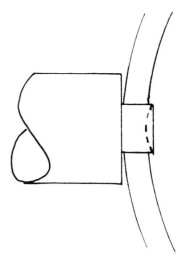

Fig 188 Making a new barrel hook.

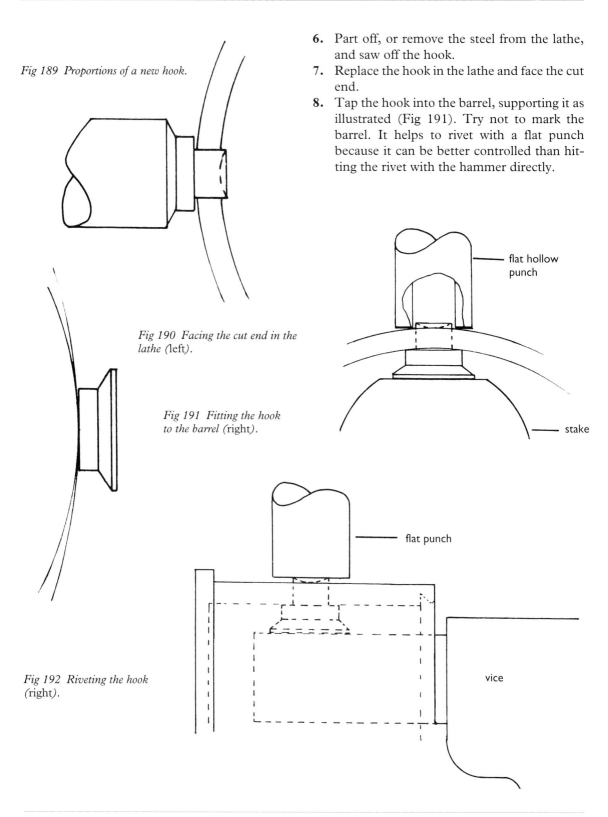

Fig 189 Proportions of a new hook.

Fig 190 Facing the cut end in the lathe (left).

Fig 191 Fitting the hook to the barrel (right).

Fig 192 Riveting the hook (right).

flat hollow punch

stake

flat punch

vice

6. Part off, or remove the steel from the lathe, and saw off the hook.
7. Replace the hook in the lathe and face the cut end.
8. Tap the hook into the barrel, supporting it as illustrated (Fig 191). Try not to mark the barrel. It helps to rivet with a flat punch because it can be better controlled than hitting the rivet with the hammer directly.

9. Finish by dressing the rivet with a file, then hold the barrel in a step chuck and grain the outside wall of the barrel with wet-and-dry or emery cloth of suitable grade to match the existing finish of the barrel. Simply polish the riveted end if the mainspring barrel is polished.

SOFT SOLDERING

There are so many examples of soft solder used inappropriately to effect repairs that in general it is frowned on. The use of soft solder is included here because there are times when it is perfectly acceptable, for example when fixing a collet onto a French strike arbor. To generalize, soft solder may be used where the original manufacturer made use of it – and there are one or two further uses I would consider legitimate.

Manufacturers have traditionally used soft solder to solder collets to arbors, and to solder barrel bottoms to barrel walls. In repair work, I would extend its use to soldering new teeth into a barrel and refacing long-case clock pallets, a practice that doesn't have universal approval but one which I justify on the grounds that at least it keeps the original pallet working in the clock.

Soft solder is a mixture of tin and lead, used in different proportions for different uses, and in particular for electrical joints and in other places where ease and convenience of use may take priority over strength. Nor does it require great heat to use. The grade suitable for horological use is $^{60}/_{40}$ – that is, 60 per cent tin and 40 per cent lead. Solder needs a flux to prevent oxidation and to help it adhere; when it incorporates a resin flux, it is known as 'resin core solder'. It can be used with a separate flux, but we do need to consider the corrosive properties of some fluxes intended for use on steel. Others are not corrosive and can be used on steel and brass. Fortunately we work almost exclusively in steel and brass, and soft solder can be used with relative ease on both.

The secret of successful soldering is preparation combined with cleanliness. A soldering iron may be used as a heat source, or heat from a flame.

Refacing Long-case Pallets
1. Take an old, unwanted piece of thin clock mainspring that will more than cover the impulse face of the pallets.
2. Mark the sides of the pallets with 'engineer's blue' and scribe freehand, or with odd legs the thickness of the spring.
3. Grind away the impulse faces to the marks, keeping the face square. Because the nibs are dead hard, they will not file.
4. Thoroughly roughen one side of the mainspring, which ultimately is to be tinned, then curve it to match the curve of the pallets. With tin snips, cut the strip into two pieces,

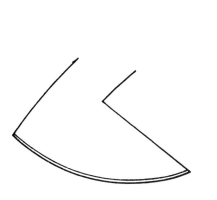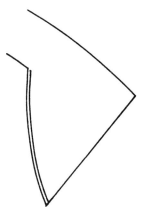

Fig 193 Long-case recoil pallets marked for capping.

one for each pallet face. An overlap of 2 or 3mm all round is desirable.

5. Spread flux over the inside roughened face of the mainspring and pre-prepared impulse faces of the pallets.

6. Heat a soldering iron, then put a little flux on the iron and a little solder. Apply the iron to the strips of mainspring, then the solder until they have a film of solder over their surface; this process we call 'tinning'. Resin-core solder may be used for this, though the presence of the resin in the solder is irrelevant to this job as it is not a good flux for soldering steel; it is just that resin-core solder is convenient to use.

7. Warm the pallets with a spirit lamp or gas flame, then after applying flux, tin the pallet faces.

8. Hold the strips in contact with the tinned pallet faces and solder together over a flame.

9. Scrub the pallets in soap and water if an acid flux was used.

10. Carefully remove the overhang of the strips. When filing or grinding, have an engaging action rather than a disengaging one, so that the strips are not torn off the pallets.

11. Finally, polish the new pallet face by first stoning with an Arkansas stone until a uniform matt grey finish is achieved, then by polishing with a clean piece of wood charged with Autosol or diamantine. A lap gives an excellent finish, but be careful that the pallets don't get caught in it or your finger could get badly cut; holding the pallets in a vice and polishing by hand is much safer. Clean all polishing material from the pallets before fitting and testing the action of the escapement.

Press in the centre of the strip while soldering to keep them flat.

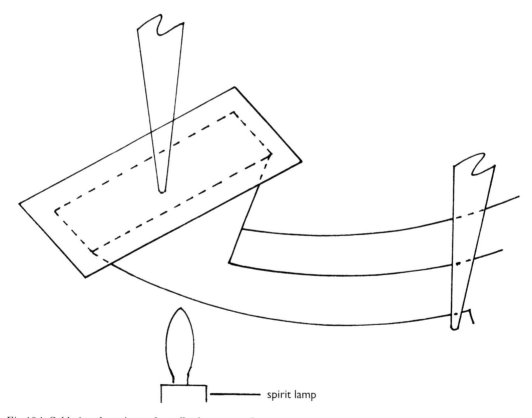

Fig 194 Soldering the strips to the pallet face over a flame.

HARD SOLDERING

Hard soldering is a process of joining two or more pieces of metal together when a joint of great strength is required and where the heat necessary to make the joint is not harmful to the metals. In clock work, the metals we are most likely to want to join are brass and steel, and for this purpose we would be likely to use 'silver solder'.

Silver solder contains copper, zinc and silver in differing proportions according to its use and is available in rods of about 1mm (⅟₁₆in) in diameter and sheets of about 50mm (2in) square. I find either satisfactory, but when using rod silver solder, I mostly flatten the rod into sheet form because it is more convenient to use that way. It is, of course, important for the solder to melt at a lower temperature than the pieces being soldered. A flux will be needed to prevent oxidization of the metal and to assist in the adhesion and flowing properties of the solder.

Hard soldering is recommended to repair a steel clock hand or to repair a broken crutch, for example.

Hard Soldering Two Lengths of Silver Steel

1. Thoroughly clean both pieces to be joined and square off the ends to be butted together as silver solder doesn't have good gap-filling properties.
2. Mix flux and water in a clean clock glass to form a thick paste that will almost, but not quite, run. Only mix enough to cover your thumb nail.
3. Cut two or three pieces of solder from a rod or sheet. You may drop one or two pieces on the floor so always cut off just a little more than you require.
4. Drop the bits of solder you have cut off into the flux.
5. Dip the end of one of the steel rods into the flux, covering only the area to be soldered.
6. With an old pair of tweezers, place a small pellet of solder close to the end of the rod.
7. With a gas torch, proceed to melt the solder. The flame is adjusted as for hardening steel but the force of the flame, and the water boiling out of the flux, will tend to throw the solder off. To prevent this, begin by stabbing with the flame until all the moisture is gone and the flux becomes molten. Even now the solder will try to work its way under the steel, so turn the piece to prevent this. Eventually the solder will curl to a ball, become shiny, and then it will run.
8. Remove the piece from the flame after the solder has run.
9. Repeat steps 5 to 8 on the other piece of rod.
10. Holding the two rods in old pliers, bring them together in the flame. Roll the ends together, then, keeping the rods touching and locking your hands together by your thumbs, as the solder melts, draw the rods back and allow the ends to butt together.

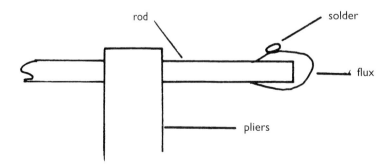

Fig 195 The rod prepared for hard soldering.

Fig 196 Butt joining the two rods. In (a) the rods are rolled around each other as the solder melts, then slid across one another for a butt join at (b).

11. Still with your hands locked together, come away from the flame and allow the joint to cool. (Cool under tap water for speed.)
12. Once cool, with safety glasses on, rest the joint on a steel block and tap it lightly with a small clock hammer. The flux will be 'glass hard' and this light hammering will remove it.
13. Clean the joint with fine emery paper.

An alternative to the above would be to clamp both fluxed rods together, apply solder, and heat until the solder flows; then cool, and clean up the piece. To test if the joint is really good, clamp one of the rods in a vice, grip the other with pliers and then bend it to and fro: the rod should break where it enters the vice, and not at the joint.

HEAT TREATMENT OF STEEL

We work with a variety of steels in various states of hardness, and apply rules of thumb when heat-treating them because for a lot of the time we are unaware of their exact nature and do not have such control over temperatures as do manufacturers.

When steel is heated it undergoes structural changes and will have varying degrees of such qualities as hardness, softness, springiness, brittleness and workability following heat treatment. These qualities are reflected by the 'colours' attained during this heat treatment, colours ranging from light straw at about 220°C, through dark straw, brown, purple, dark blue (300°C), light blue and grey, then changing to a blood red

and eventually to cherry red at 800°C and then to yellow and white hot.

The processes we are chiefly concerned with are hardening, tempering and softening. Hardening is a process which precedes tempering; tempering means leaving the steel in a state of hardness for the work it has to do; and softening is for fashioning, though it is still possible to fashion steel, to a point, even when it is tempered to dark blue. Most of the time steel is fashioned in its soft state, for example when manufacturing new arbors and pinions. For the job it has to do, the steel should be brought to a particular degree of hardness and toughness, and when the required degree of hardness is harder than its present state, the steel always needs hardening dead hard first.

Hardening Steel

Bring the steel to a cherry red (about 800°C). At this point, briefly, the temperature of the steel will not rise but instead, the steel will undergo a structural change. There is no way of knowing when the change is complete, but holding the steel in the flame at cherry red for a couple of seconds will allow the change to come about. Now quench the steel in water at room temperature. The success of the change depends on the speed of cooling, so have the coolant close to the flame so that there is no appreciable drop in temperature while the steel is being transferred. The hardening needs to be done in shaded conditions, that is, without sunlight or a bright light shining on the steel. This is because cherry red in the shade would be only blood red in sunlight, and cherry red would be achieved at a much higher temperature.

Some steel parts, for example some gravers, crack when quenched directly into water. This doesn't happen if a little oil is floated on the water, so that quenching is through oil into water.

If what you are quenching is long and thin, like an arbor, quench by immersing vertically and not horizontally, because this will minimize the likelihood of the steel bending. Some scaling will occur during hardening but this is unavoidable to a large degree. It is claimed that quenching into salt water minimizes scaling, but my experience is that it doesn't stop it.

At times I have covered steel in Easy-flo hard soldering flux before heating to cherry red to minimize scaling, and it does seem to give a better result.

The steel should now be 'dead hard' but it will be brittle, so we have to remove the brittleness and introduce toughness.

Tempering Steel

After tip hardening a punch, say, the steel will be grey where it became hot beyond about 300°C, and other colours between the bright metal and the heat-treated tip. These colours are oxidization colours and are used to determine the degree of temper after hardening. Once the metal has gone through particular colour changes, for example when hardening, the metal has to be brightened again to see the colours to judge the degree of temper. For this you could use an Arkansas stone, an emery cloth, a buff stick, or an abrasive block. I have used rust remover to remove oxidization colours, when re-pivoting for example (though scrub in soap and water if rust remover is used because it is corrosive). Not all the above suit all circumstances, however.

To temper, bring the steel to the appropriate colour, then remove the heat source and allow the steel to cool. As an alternative, quench in water. If it is a graver that is being quenched, there is no need to go through oil into water because the shock is not sufficient to crack it. When tempering uniformly, that is, throughout a whole piece, there is no need to quench: simply removing the steel from the heat source is sufficient. When tempering a punch or graver with the flame playing directly on the steel above the point of interest, you do need to quench to avoid too much heat travelling to the tip, thereby tempering beyond the required colour. Under these conditions, the steel itself becomes the heat source and the heat will continue to flow unless the steel is quenched.

Below is a tempering guide for items we often have to heat-treat:

Light straw:	Graver, lathe tools and drills
Brown:	Taps
Brown to purple:	Stakes and punches
Dark blue:	Springs, arbors and screws

If you temper beyond the required colour, you must re-harden before tempering again.

Softening Steel

When you wish to fashion something out of steel that has already been hardened and tempered, you will have to soften it first. There are exceptions, of course: for example 'blue steel', and ordinary steel that has been hardened and tempered to blue, will turn with a graver but not with a tool bit in a centre lathe. Blue steel can be filed and drilled with a tungsten carbide drill or spade drill, but is not easily drilled with an HSS drill.

To soften steel for 'working', heat it to a cherry red, then allow it to cool slowly. Often it is very difficult to draw the temper of previously hardened and tempered, old steel. Fortunately, mostly when this is necessary it is for repivoting an arbor, and a tungsten carbide drill overcomes the need to soften.

DECORATIVE AND PROTECTIVE FINISHING

Screws and hands are blued to give them a finish which is both decorative and protective. Hands in particular are blued without hardening first because only the decorative and protective aspects of the heat treatment are of interest.

To blue clock hands, take a copper or brass tray with a handle and hold it in a vice, and fill it with clean, fine brass filings. Place the hand in these so that only its top surface remains uncovered, and play a flame under the tray so that heat is transferred to the hand. As soon as it becomes a uniform dark blue, simply remove it from the filings and allow it to cool.

A hand is prepared for bluing by polishing its top with polishing paper, Autosol or diamantine, then cleaning it and placing it in the brass filings without touching its top.

A screw may be blued in alternative ways: either hold it by the thread with an old pair of tweezers and blue it with a spirit lamp; or place it in a block of brass so that the thread enters a drilled hole and the underneath side of the screw is resting on the block. Next, heat the brass block until the screw turns to dark blue, then immediately turn the block over so that the screw drops onto a suitable surface such as a tin tray.

HARDENING AND SOFTENING BRASS

Brass is supplied in a reasonable state of hardness already, and for making pinions, cocks, bridges and plates, no further hardening is necessary. However, if you needed to make a click spring out of brass, some further hardening will be necessary.

Hardening Brass

Brass can be, and is, hardened in a variety of ways, all of which involve compression, either drawing through a die, rolling, hammering or burnishing. Often in the workshop you will need to use one of the last two methods to harden brass.

To make a brass click spring for a long-case clock, first select a piece of suitable brass a little oversize, then hammer it to compress it, making it more 'springy'. File, or perhaps draw-file, drill and shape the click to match the original, and fit it to the great wheel.

After silver soldering a brass click, hammering to restore hardness would be out of the question, so 'skin'-harden by burnishing. For this, rub an oiled burnisher firmly over all the faces of the click, leaving a hard skin on the outside though a soft centre. This should restore the springiness of the click so that it works efficiently again.

Softening Brass

Simply heat the brass to a dull red and quench in water or allow to cool slowly. The advantage of quenching is that you can begin work right away without waiting for the brass to cool. Brass doesn't harden by quenching in the same way as steel.

When pivoting, avoid softening brass wheels. They can't be re-hardened, and it is better not to allow the problem to arise. If you need to soften an arbor to pick up a centre for drilling and you are working close to a wheel, consider wrapping a wet paper towel around the brass wheel to prevent heat from travelling into the brass. Alternatively you could cover the wheel with a paste designed to stop heat spreading.

14 GEAR CUTTING

There was a time when gear cutting was considered too specialized for the enthusiast, but more and more hobbyists are becoming involved, partly because of the short courses on gear cutting now offered, but also because of the availability and comparatively low cost of gear-cutting equipment. Some may wonder why wheel teeth and pinion leaves are the shape they are, because it would seem that straightforward 'spoke-like' shapes ought to work just as well. The problem with such a shape is that, assuming the driving wheel turned at a uniform rate, the wheel it drives would have an erratic rate. Thus in Fig 197a, for a particular small angle of turn of the driving wheel, a similar angle will be turned by the driven wheel. Equally it might be clear that for a fairly large turn of the same driving wheel, the driven wheel will turn but very little (Fig 197b).

In horology, the gearing is designed so that there is a constant ratio between a pair of mating gears, normally wheel and pinion. In addition, we want our gears to operate dry, and to stop and start perhaps 18,000 times an hour. If a good impulse is to be given, once unlocked, the wheels and pinion should have minimal inertia and accelerate quickly. For most horological clock work, our gears are designed to give a step up in gear ratio from a slow-moving great wheel to a relatively fast-moving escape wheel.

As we know, a watchmaker's 8mm lathe is belt-driven and a small amount of belt slip could be tolerated and probably not even noticed. If, however, we wanted to screw-cut on the lathe, it would be a different matter because for this we need a gear-box to give uniform lead. In the same way our clock needs gear teeth for wheels to drive pinions, although for the purpose of studying gears, we can think of the wheels as cylinders rolling together.

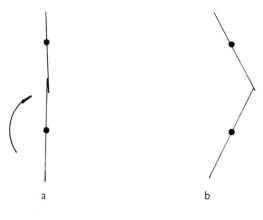

Fig 197 In (a), at the start of the action, the driving wheel will turn at about the same rate as the driven wheel. In (b) a large turn of the driving wheel turns the driven wheel very little.

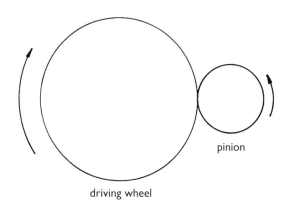

Fig 198 Two rolling circles representing a driving wheel and pinion.

In real gears, there are theoretical pitch circles that have the same diameter as our representative rolling circles. On these circles teeth are formed, part of which are above the pitch circles and part below. That part of a tooth above the pitch circle is known as the addendum, and the part below it is the dedendum.

It is important that you are familiar with the following terms: addendum, dedendum, pitch circle (PC), pitch circle diameter (PCD), pitch circle radius (PCR) and distance of centre

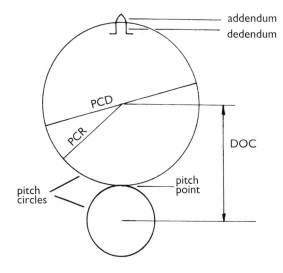

Fig 199 Terms with which you need to be familiar.

(DOC). When a pair of gears is correctly meshed, the line of action between a tooth and pinion leaf will pass through the pitch point, that is, where the two circles meet.

Horological gearing is often described as 'cycloidal' or 'circular arc', which means that the acting faces of the teeth are arcs of circles approximating to epicycloidal and hypocycloidal curves. This is to ensure that we have the condition of uniform lead. Where great strength is necessary, for example in winding work, approximations to involute curves are used.

A cycloid is 'the path described by a point on the circumference of a circle as the circle is rolled without slipping along a straight line'.

An epicycloidal curve is 'the path described by a point on the circumference of a circle as it is rolled without slipping around the outside of another circle'.

A hypocycloidal curve is 'the path described by the point on the circumference of a circle as it is rolled without slipping around the inside of another circle'.

The rolling circle is known as the generating circle, and the circle it rolls around as the base circle. If the diameter of the generating circle is less than half the diameter of the base circle, the result is a convex curve. If the diameter of the generating circle is greater than half the diameter of the base circle, a concave curve is formed. Of more significance to the horologist is the 'curve' formed when the diameter of the circle is exactly

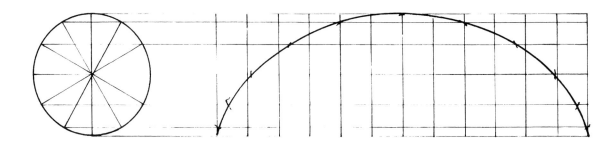

Fig 200 A cycloidal curve.

half the diameter of the base circle, because a straight line is produced.

An involute curve is the path described by a point on a piece of taut string as it is unwound from a cylinder. It is the curve seen on a French strike snail.

To summarize, the acting faces of the wheels are arcs of circles closely representing these curves. The tips of the teeth are not used but offer 'tip relief', that is, they prevent the teeth from digging into a pinion in the same way as skis have a turned-up end to allow them to ride over irregularities of the snow.

GEAR CUTTERS

Cutters are selected according to their 'module' (M). Module is a metric measurement and is defined as 'the number of millimetres of a pitch-circle diameter per tooth'. A formula for calculating module is:

$$M = \frac{PCD}{\text{wheel teeth or pinion leaves}}$$

Now this is not very helpful in a practical situation, because the PCD is somewhere between the tip and the root of a tooth but we are not sure just where. (There are other formulas that could also be used, but they are not helpful in the same way.) A better formula to remember because it can be used is this :

$$M = \frac{DOC \times 2}{\text{sum of the teeth in the wheel and leaves in the pinion}}$$

$$\text{written } M = \frac{DOC \times 2}{t + 1}$$

Remember, it is the leaves in the pinion driven by the wheel that are being calculated, not the wheel's own pinion! It is the same module, but a different cutter for a mating wheel and pinion.

From the formula, it might be thought that there would be a vast number of modules to cover, but in fact the difference between modules is so small that they have been rationalized. For most clock work, the choices for modules can be rounded to select a cutter from the following range:

0.4	0.7
0.45	0.75
0.5	0.8
0.55	0.85
0.6	0.9
0.65	0.95
	1.00

To obtain a particular wheel cutter you will need to state details of the wheel cutter, including the module, square or round bottom tooth, bore (the hole size in the cutter, which is mostly 7mm) and HSS (this is high-speed steel: perhaps it will be all that is available). For a pinion cutter you will need to state the pinion cutter, module, number of leaves and bore (again, probably 7mm, although it could be smaller for a smaller module. It is likely that you will have to accommodate the size they are manufactured in).

Once you have the cutter, enter it into the teeth or leaves as appropriate to compare the profile. One of the problems with cutting a wheel or pinion for an old clock is that the modern standards used to make a cutter are different from the old standards. To overcome this problem we could either make a cutter; or cut a gear with the nearest module cutter and modify the tooth or leaf by hand; or cut both wheel and pinion to obtain a match.

Before leaving the question of cutter selection, I will mention that there is an Imperial system, known as diametrical pitch, or DP. I do not recommend its use, but for those who wish to know, DP is the number of teeth in each inch of pitch diameter.

$$M = \frac{25.4}{DP} \qquad DP = \frac{25.4}{M}$$

The horological industry works almost exclusively in module and I see no virtue in the Imperial system for cutting horological gears.

Calculating the Outside Diameter of Blanks

As already mentioned, cutting new wheels and pinions for old clocks can be a problem, and there are times when the original wheel's outside diameter (OD) is more appropriate than using a theoretical OD. The theoretical OD for one cutter manufacturer, P.P.Thornton (Successors) Ltd, is this:

$$OD = M(N + \text{addendum allowance})$$

where N is the number of teeth or leaves (addendum correction is taken from a set of tables made available by PP Thornton (Successors) Ltd, *see* Useful Addresses).

Materials

Recommended for pinions is KEA 108 which is a free cutting steel and is hardenable for those who wish to heat-treat arbors and pinions after manufacture. A good cutting speed is 120 to 260 revolutions per minute (RPM).

For steel up to 5mm (³⁄₁₆in) in diameter, oil hardening is recommended; for steel over 5mm (³⁄₁₆in), use water hardening. Arbors and pinions should be tempered to dark blue. For wheels, CZ 120 is recommended with cutting speeds of between 1,000 RPM for 25mm (1in) diameter brass and 1,850 RPM for 13mm (½in) brass. Fly cutting is too advanced for this book, but for those already experienced, a speed of 8,000 RPM would be about right for wheels.

MAKING PINION BLANKS

When making pinion blanks, we have to consider the following as priorities: maintaining concentricity between pivots, pinions and eventually the wheel, we have to hold our blank in the lathe collet probably with support on the other end of the arbor; and we must also have room for the cutter to approach the blank, cut the leaves and clear the blank.

As an example, let us assume we are to renew an arbor and pinion for a long-case clock third wheel in a Myford lathe. The cutter will be held on a mandrel in the lathe headstock, and the blank is

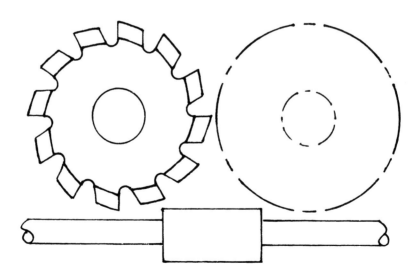

Fig 201 For a full cut, the cutter needs to enter the pinion before the centre line of the cutter passes the end of the pinion and cut beyond the end of the pinion.

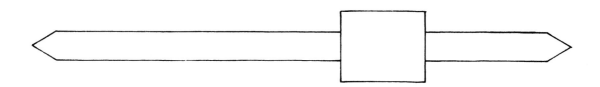

Fig 202 A pinion blank.

held in a dividing head. Eventually we will need to decide whether the cutter will be under the blank or over it; for the moment, however, let us consider what will be happening to the cutter and pinion blank. If we decide to put the cutter over the blank, the profile can be inspected more easily. This has the advantage of raising the blank into the cutter, avoiding damaging backlash in the vertical slide's lead screw.

The rotating cutter will be fixed in the headstock and will be about 25mm (1in) in diameter. To get full teeth and spaces so that it will not matter where the wheel engages in the pinion, the cutter has to enter the pinion so that the centre line of the cutter is just outside of the start of the pinion. The space will be cut and the cutter centre line must just clear the end of the pinion. This is made more clear in Fig 201.

To satisfy the action between cutter and blank, the blank may need to be made over-long.

Making the Blank
1. Sketch and dimension the intended blank, allowing room for the cutter.
2. Select a piece of KEA 108, the nearest fractional size over the diameter of the pinion to be cut that will also fit your lathe collet.
3. Turn male centres on each end of two blanks (if you have a 'spoil' you will be glad you made a spare blank), with 60° included angles to fit the centres in the lathe which will be used for finishing the arbor. Stone the male centres with an India or Arkansas stone.

4. Turn your blanks so that the pinion OD is the same size as your original pinion, and the diameter of that part of the arbor that is to be held in the collet is to the nearest fractional size (oversize) to suit an available collet.
5. You should now have two blanks looking similar to that depicted in Fig 202.

Setting up for Pinion Cutting
1. Remove the top slide and toolpost from the cross slide.
2. Mount the vertical slide in the nearest 'T' slot of the cross slide.
3. Secure the dividing attachment to the vertical slide, and with a test bar in a collet in the dividing attachment, check that the bar is horizontal and at right-angles to the bed. Use a dial test indicator (DTI) for this.
4. Check that the division plate in the dividing attachment is the correct one, then set the sector arms. Here we will be cutting an eight-leaf pinion on a 60:1 ratio, which means sixty turns of the handle to one revolution of the collet: $^{60}/_{8}$ cquals $7\frac{1}{8}$, therefore we need to turn the handle by seven turns and $\frac{1}{8}$ of a turn. If the division plate has a circle of thirty-two holes, then $\frac{1}{8}$ of thirty-two is sixteen, so set the sector arms so that with one arm resting against the pin in the division plate, the other arm is set to show sixteen empty holes between the arms; that is, a total of seventeen holes.
5. Put the cutter arbor in the headstock with about 35mm (1⅜inch) overhang, which

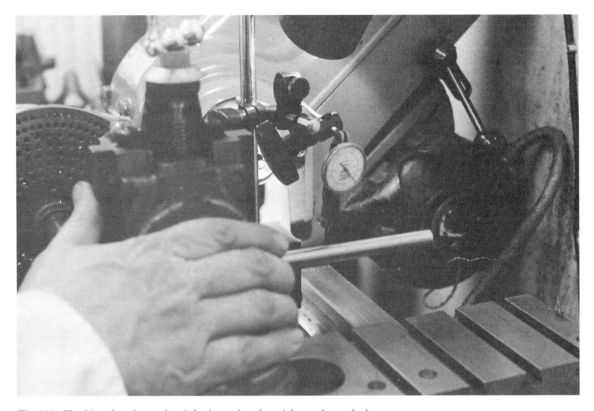

Fig 203 *Checking that the test bar is horizontal and at right-angles to the bar.*

Fig 204 *Setting the sector arms.*

should ensure that as work progresses, the two nose pieces don't touch. Clean the shoulder to take the cutter and the cutter itself.

6. Mount the cutter on the arbor so that it cuts with the lathe turning in its more usual direction. This way, the centre in the overarm will ensure that the blank doesn't move because of the direction of the cut. If the cutter turned in the opposite direction, it could push the arbor back into the collet.

Now we are going to make sure that the cutter cuts on the centre line of the blank.

7. Put a true pointed piece of steel in a collet in the dividing attachment then, checking with an eyeglass, move the saddle so that the pointed bit of steel is in the centre of the cutter.

8. Once the centre is found, by this or other means, lock the saddle and make sure that the lead screw is disengaged. Remove the pointed piece of steel you have been using to centre the blank.

9. Secure the blank so that the colleted end protrudes from the collet by a little more than half the diameter of the cutter.

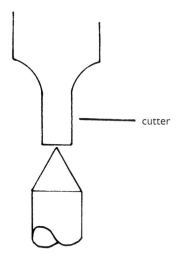

Fig 205 Centering the dividing attachment under the cutter.

10. Adjust the overarm so that the blank is held firmly. The overarm must not strain the blank by pulling it over.

11. Adjust the speed of the cutter to 210 RPM.

12. Put engineer's blue (or red) round the blank so that later, when cutting the spaces, a 'witness' at the top of a leaf is seen more easily.

13. By working between the vertical slide and the cross-slide, position the blank with respect to the cutter so that a full depth of cut is not possible, yet a slight rounding of the top of a leaf is likely. At the same time, make sure that the tip of the cutter clears the arbor.

14. Turn the headstock one revolution, by hand, just to confirm that all is well.

Cutting the Pinion

1. Wearing safety glasses, turn the lathe on and take a cut through the pinion blank; keep it moving to avoid creating a hard spot on it. Turn the lead screw slowly and steadily. If there is another person available to help, get him or her to apply a little cutting oil to the piece with pegwood as work progresses; it will prolong the life of the cutter and give a better finish. If cutting lubricant is brushed on, don't get the bristles caught in the cutter.

2. When a space has been cut, return the blank to your starting point and switch off the lathe.

3. Inspect the cut: if you can see a little rounding at the top of the blank, proceed to step 4; if not, turn the lathe on and bring the blank up into the cutter by 0.010inch and take a further cut. Keep this up until a slight rounding at the top of the space is seen.

4. Once rounding is seen, as long as the cutter is free of the blank, rotate the plunger in the dividing attachment by seven complete turns plus half a turn. Once the plunger is engaged in its hole, rotate the sector arms by half a turn so that the arm is resting against the pin. Both plunger and sector arms are turned clockwise.

5. Take a further pass through the blank and back again.

6. Turn the lathe off and inspect: you now have two spaces and one leaf. Inspect the top of

the leaf: if you have a wide band of marking ink at the top of a tooth, switch the lathe on, raise the blank by a further 0.010in or less (not more), and take a further pass through the same space: return the blank to the starting point.

7. Provided the cutter is still free of the arbor, rotate the plunger a further seven-and-a-half turns clockwise and after engaging the plunger, rotate the sector arm. Always work to the same pattern as then there is less likelihood of a mistake. If you have to turn the blank backwards, always go back further than necessary, then forwards again; this is to take up any slack in the worm and pinion.

8. Take a pass through the blank and back again which will leave you with two partly formed leaves.

9. Inspect the top of the leaf you have just cut, and assess the width of the marking ink at the top of the leaf.

Please note: Once rounding of the teeth starts, each time you raise the blank by the same amount, the rate at which the width of the marker ink narrows, accelerates. Often, the cutter enters not only the pinion blank but also the arbor itself. This is quite common, but it does mean that to advance the blank to the next space, the blank has to be lowered to clear the cutter.

Fig 206 Often the cutter enters the arbor itself.

Always check that the cutter is clear of the blank before turning the blank.

10. Keep making passes in the way you have already been doing but raise the blank less each time, until you have a very narrow strip of marking ink along the top of one leaf of about 0.002inch. Now, at this depth, go all around your blank.
11. Finally, raise the blank by 0.001inch and take a light finishing cut through all spaces.
12. Remove the blank from the lathe, inspect for upright leaves, and finish by turning the pivots between centres and generally finishing the blank.

If the wheel is mounted directly onto the pinion, a shoulder is turned with a slight undercut so that the wheel lies flat and a steep undercut is produced to rivet the wheel. With a thin leaf I sometimes find that the leaves can start to lean whilst turning with a lathe tool or graver. To prevent this from happening, fill the spaces between leaves with sealing wax, which will support the leaves while turning. Eventually, dig off the wax and wipe over any that remains with methylated spirit on a cloth.

As already mentioned, in high quality work, arbors are hardened and tempered to dark blue. After hardening and tempering, polish the pinion by shaping a piece of wood – for example from an old buff stick – to fit a pinion space, then with polishing paper wrapped around the end of the buff stick, work it up and down each pinion space. As a final finish, shape a piece of wood as before but charge it with Autosol or diamantine and work that up and down the pinion leaves. The arbor itself can be cleaned up in much the same way.

MAKING A WHEEL BLANK

The same formula is used for the OD of a wheel as for the OD of a pinion.

1. Start with a sheet of CZ 120, the thickness of which is the same as the wheel to be cut.

Scribe a circle not less than the OD of the blank you require.
2. With a piercing saw or a round blade, tyle-cutting saw, cut out the blank close to the scribed line.
3. Round any sharp corners on the blank with a file.
4. Centre pop and drill a hole suitable for reaming to fit your wheel mandrel. Put a backing washer to support the wheel during cutting onto the mandrel with the blank.
5. Secure the mandrel in the lathe headstock, and turn the blank OD exactly to size.
6. Remove the mandrel and set the wheel cutter up in the headstock so that it cuts with the lathe spindle turning in the reverse directive to normal.
7. Centre the collet in the dividing attachment with respect to the cutter in a similar fashion to the way it was centred for pinion cutting. Lock the saddle and adjust the speed of the lathe to the fastest practicable speed.
8. Put the blank in the dividing attachment with marking ink around the circumference, and set the sector arms according to the number of teeth and the ring of holes in the division plate.
9. Adjust the depth of cut on the vertical slide and proceed to cut the teeth in a similar way to that described for pinion cutting. As for pinions, the teeth may be cut in one pass with a light finishing cut before removing from the lathe. Start your cut with the cutter behind the blank so that the blank is advanced to the cutter – that is, with the blank moving away from you.

If two or more identical wheels are wanted, blanks may be stacked on the mandrel and several wheels can be cut simultaneously.

When you have finished, the wheels are 'crossed out' – a term used when referring to marking out the arms or spokes – and drilled, and then the portion to be removed from between the arms can be sawn out with a piercing saw.

The arms and rim are finished with a crossing file then, for quality work, the crossings are burnished with an oiled darning needle. Triangular sailmaker's needles will allow you to get tighter into the corners of the crossings.

To deburr the wheel, glue polishing paper to small squares of melamine and rub the wheels over the paper.

SECURING A WHEEL TO A PINION

1. Make or select a flat hollow punch to drive the wheel onto the pinion.
2. Make or select a round-ended hollow punch to rivet the wheel supporting on a brass stake to save damaging the pinion.
3. Spread the rivet, then change to a flat hollow punch to flatten it.
4. When you have finished, grip the arbor in the brass jaws of a vice, and check that the wheel is secure. In high class work, the rivet may be

polished but for mass-produced clocks, the finish may be that left by the punch.

5. Check that the wheel is true.

SECURING A WHEEL TO A BRASS COLLET

Let us assume that a wheel is to be fastened to an arbor with a badly damaged collet on a French clock.

1. Make a sketch of the finished job with dimensions so that the wheel and collet can be correctly positioned without interfering with any other parts of the clock.
2. Warm the old collet over a spirit lamp or gas flame to melt the solder that secures the collet, and slide it off the arbor. If there is any difficulty with this, just turn the old collet off.

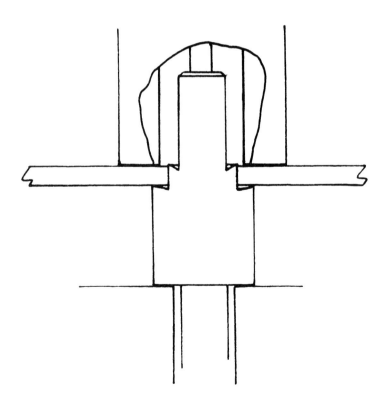

Fig 207 Driving the wheel onto the pinion with a flat, hollow punch.

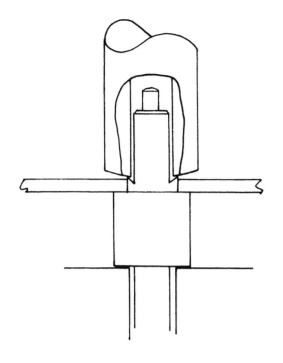

Fig 208 Riveting the wheel.

3. Make a new rough collet and soft solder it onto the arbor. At this stage, the collet is just a cylinder soft soldered to the arbor.

4. Because it is important for the pivots, pinion and wheel to be concentric with one another, you should turn the shoulder and diameter to take the wheel between centres. For this you will need a carrier chuck with a female centre, carrier and a tailstock with a female centre.

5. Turn the collet with a slight undercut for the wheel to rest against and a steep undercut for the rivet. The diameter for the wheel should be such that the wheel has to be tapped on.

6. Turn the back of the collet roughly to shape either by reversing the arbor in the lathe or by turning it in the position it is already in.

7. Rivet the wheel to the collet. To do such jobs I have made a support for the collet from a couple of pieces of brass shaped as illustrated in Fig 212. New holes are drilled as they are needed.

To use, the cheeks are placed over the arbor and collet, then the cheeks are held in the vice.

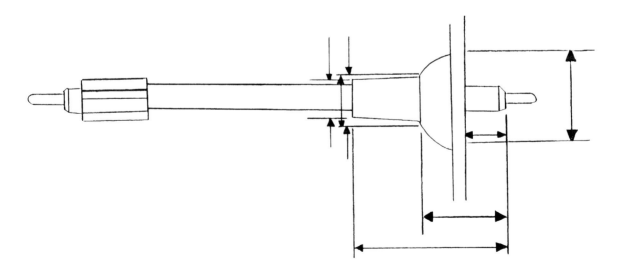

Fig 209 Sketch of the finished wheel with dimension lines. Enter your own dimensions.

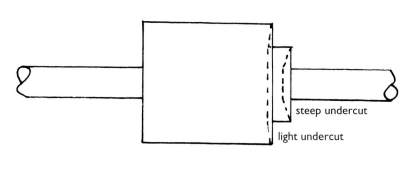

Fig 210 A light and a steep undercut for riveting.

steep undercut

light undercut

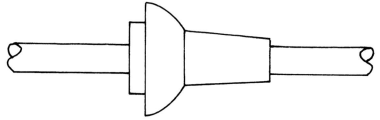

Fig 211 The back of the collet roughed out.

Fig 212 Home-made brass cheeks for riveting a wheel.

With the cheeks giving support under the collet, rivet in the same way as explained for riveting a wheel to a pinion.

8. After riveting, put the arbor back in the lathe and tidy up the collet.
9. True the wheel.

FITTING TEETH

Currently there is less need to fit teeth to barrels and other wheels than heretofore because it has become easier to cut gears for yourself and to make use of the increasing number of people offering a gear-cutting service. However, in the interest of conservation, fitting individual or groups of teeth may be a preferred option.

There are three tried and trusted ways of fitting teeth. **Method one** is suitable for up to three teeth, particularly in a barrel where the bottom of the barrel is thinner than the height of the teeth. **Method two** is used when more than three teeth have broken out; and **method three** could be used on a fusée wheel where there is sufficient thickness of the wheel and enough depth to where the relief is turned to accommodate the ratchet wheel, click and click spring.

Method One
1. Remove the barrel arbor, cover and spring.
2. Upright any bent teeth using pressure with a knife against the root of an adjacent tooth and the tip of the bent tooth.
3. Using a hacksaw with a saw blade ground if necessary to a tooth thickness, slit down the face of the barrel by about twice the height of the teeth without cutting right through, or up the wall. The slit must be in the right position but may be a little wider than a tooth.
4. Deburr the slits.
5. File a bit of hard brass to an interference fit in the slits already made. Allow a small overhang on the face of the barrel and above the tooth height for holding.
6. Tin the prepared strip on the faces that mate with the barrel.
7. Apply flux to the slits, and push the strips into position.
8. Chip a few small lumps of solder from your soldering stick and place them beside the strips for gap filling.
9. Hold the barrel over a spirit lamp or a gas flame so that the flame plays on the inside of the barrel, and heat until the solder runs.

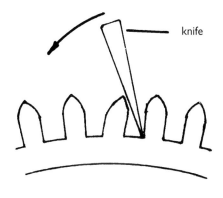

Fig 213 Uprighting a bent tooth.

Fig 214 The barrel is slit with a hacksaw blade.

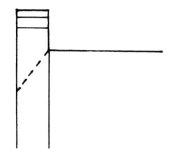

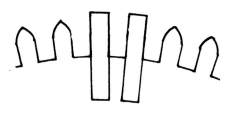

Fig 215 Hard brass strips fitted to a barrel with overhangs.

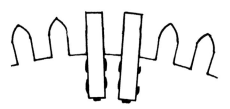

Fig 216 Chips of solder placed for soldering.

Avoid overheating because the barrel bottom may well be soft soldered to the wall and we don't wish them to part. Apply your heat locally. Further, we don't wish to soften the brass.

10. Remove the surplus brass with a piercing saw and file the teeth to shape. Notice the 'strips' have now become 'teeth'. Check there is no evidence of your repair on the inside of the

barrel, and if there is, remove any obstruction to the mainspring.

11. Replace the assembled barrel in the clock with the wheel it drives, and check that the new teeth pass freely. Hopefully, if you were blindfolded you couldn't find your new teeth by sound or feel!

12. Remove the barrel, and finish to its original finish.

Method Two

1. From a scrap barrel similar to the one being repaired, cut out the number of teeth to be replaced into your barrel.

2. Dovetail the original barrel and shape your insert to be an interference fit into the dovetail.

3. Tin the surfaces to be soldered, then insert the new teeth and solder as above. It may be necessary to turn the inside of the barrel with a boring tool to prevent the mainspring binding.

Fig 217 Dovetail your barrel and new teeth.

Method Three

1. Select two screws with a core diameter of about the same as a tooth thickness or slightly larger.
2. Where a tooth has broken out, centre pop and drill two holes for tapping to a depth of about double the height of a tooth.
3. Tap both holes with a plug tap.

4. Screw one screw into one of the holes so that it bottoms and is tight. Special glue could be used here, although I have never found it necessary.
5. With a piercing saw, remove the head of the screw and surplus thread, and shape the screw to a tooth profile.
6. Repeat with the other screw.
7. Test your new tooth with its mate.

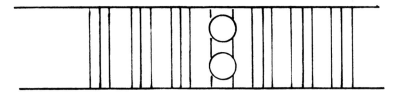

Fig 218 Drill two holes in your barrel for tapping.

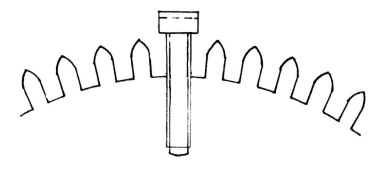

Fig 219 One of the screws inserted.

USEFUL ADDRESSES

The British Horological Institute
Upton Hall
Upton
Newark
Nottinghamshire
NG23 5TE

PP Thornton (Successors) Ltd
The Old Bakehouse
Upper Tysoe
Warwickshire
CV35 0TR

TOOL AND MATERIAL SUPPLIERS

AG Thomas
Tompion House
Heaton Road
Bradford
West Yorkshire

HS Walsh Son Ltd
1/2 Warstone Mews
Warstone Lane
Birmingham

Mahoney Associates
PO Box 254
58 Stapleton Road
Bristol
Avon

Charles Greville & Co
224 The Mill
Millers Dale
Buxton
Derbyshire

GK Hadfield
Rock Farm
Chilcote
Swadlincote
Derbyshire

HS Walsh & Sons Ltd
243 High Street
Beckenham
Kent

Meadows & Passmore
Farningham Road
Crowborough
East Sussex

Southern Watch & Clock Supplies Ltd
Precista House
48/56 High Street
Orpington
Kent

E Cousins
Unit J, Chesham Close
Romford
Essex

INDEX